Communications
in Computer and Inf

212

Tzuu-Hseng S. Li   Kuo-Yang Tu
Ching-Chih Tsai   Chen-Chien Hsu
Chien-Cheng Tseng   Prahlad Vadakkepat
Jacky Baltes   John Anderson
Ching-Chang Wong   Norbert Jesse
Chung-Hsien Kuo   Haw-Ching Yang (Eds.)

Next Wave
in Robotics

14th FIRA RoboWorld Congress, FIRA 2011
Kaohsiung, Taiwan, August 26-30, 2011
Proceedings

 Springer

Volume Editors

Tzuu-Hseng S. Li, E-mail: thsli@mail.ncku.edu.tw

Kuo-Yang Tu, E-mail: tuky@nkfust.edu.tw

Ching-Chih Tsai, E-mail: cctsai@dragon.nchu.edu.tw

Chen-Chien Hsu, E-mail: jhsu@ntnu.edu.tw

Chien-Cheng Tseng, E-mail: tcc@ccms.nkfust.edu.tw

Prahlad Vadakkepat, E-mail: prahlad@nus.edu.sg

Jacky Baltes, E-mail: jacky.baltes@gmail.com

John Anderson, E-mail: andersj@cs.umanitoba.ca

Ching-Chang Wong, E-mail: wong@ee.tku.edu.tw

Norbert Jesse, E-mail: norbert.jesse@udo.edu

Chung-Hsien Kuo, E-mail: chkuo@mail.ntust.edu.tw

Haw-Ching Yang, E-mail: hao@nkfust.edu.tw

ISSN 1865-0929   e-ISSN 1865-0937
ISBN 978-3-642-23146-9   e-ISBN 978-3-642-23147-6
DOI 10.1007/978-3-642-23147-6
Springer Heidelberg Dordrecht London New York

Library of Congress Control Number: 2011933832

CR Subject Classification (1998): I.2.9, I.2, I.4, J.2

Typesetting: Camera-ready by author, data conversion by Scientific Publishing Services, Chennai, India

Printed on acid-free paper

Springer is part of Springer Science+Business Media (www.springer.com)

Preface

This volume contains papers selected from the 14th FIRA RoboWorld Congress, held in Kaohsiung, Taiwan, ROC, during August 26–30, 2011.

The Federation of International Robosoccer Association (FIRA - www.fira.net) is a non-profit organization, which organizes robotic competitions and meetings around the globe annually. The Robot Soccer competitions started in 1996 and FIRA was established on June 5, 1997. In FIRA 2002, held in Seoul, Korea, the HuroCup competition was added, which focuses on developing full-capability humanoid robots similar to humans. The Robotics competitions are aimed at promoting the spirit of science and technology to the younger generation. The congress is a forum in which to share ideas and the future directions of technologies, and to enlarge the human networks in the robotics area.

The objectives of the FIRA Cup and Congress are to explore technical developments and achievements in the field of robotics, and to provide participants with a robot festival including technical presentations, robot soccer competitions, and exhibitions.

The main theme of the 16th FIRA RoboWorld Cup was "Enjoy Robot and Enjoy Life." Under this slogan, three international conferences were held for greater impact and scientific exchange:

- 8th International Conference on Computational Intelligence, Robotics and Autonomous Systems (CIRAS)
- Third International Conference on Advanced Humanoid Robotics Research (ICAHRR)
- Third International Conference on Education and Entertainment Robotics (ICEER)

This volume consists of selected quality papers from the three conferences. The volume is intended to provide readers with the recent technical progress in robotics, human-robot interactions, cooperative robotics and related fields. The volume has 34 papers from the 110 contributed papers at the 14th FIRA RoboWorld Congress.

The editors hope that this volume is informative to the readers. We thank Springer for undertaking the publication of this volume.

Tzuu-Hseng S. Li
Kuo-Yang Tu

Industrial Forum and Exhibit Co-chairs

Chung-Hsien Kuo   National Taiwan University of Science and
                         Technology, Taiwan
Hans Pao   Innovati, Inc.
Anthony Lu   Institute of Intelligent Robots, Taiwan
                         Development Institute, Taiwan
Chia-Yen Chen   Precision Machinery Research Development
                         Center, Taiwan
Jui-Chi Chen   Precision Machinery Research Development
                         Center, Taiwan

International Relations Chair

Kao-Shing Hwang   National Chung Cheng University, Taiwan

Invited Session Co-chairs

Kuan-Yow Lian   National Taipei University of Technology,
                         Taiwan
Shui-Chun Lin   National Chin-Yi University of Technology,
                         Taiwan
Ming-Yang Cheng   National Cheng Kung University, Taiwan

Venue Co-chairs

Ming-Yuan Shieh   Southern Taiwan University, Taiwan
Kuo-Ching Tseng   National Kaohsiung First University of Science
                         and Technology, Taiwan

Award Co-chairs

Jyh-Cheng Yu   National Kaohsiung First University of Science
                         and Technology, Taiwan
Chao-Lin Kuo   National Kaohsiung Marine University, Taiwan

Financial Chair

Ching-Kao Chang   National Kaohsiung First University of Science
                         and Technology, Taiwan

Organization

Advisory Board

Tsu-Tian Lee — National Taipei University of Technology, Taiwan

Han-Pang Huang — National Taiwan University, Taiwan

Jong Hwan Kim — Korea Advanced Institute of Science and Technology, Korea

Peter Kopacek — Vienna University of Technology, Austria

Roger C.Y. Chen — National Kaohsiung First University of Science and Technology, Taiwan

Jyh-Horng Chou — National Kaohsiung First University of Science and Technology, Taiwan

Rong-Fong Fung — National Kaohsiung First University of Science and Technology, Taiwan

General Co-chairs

Tzuu-Hseng S. Li — National Cheng Kung University, Taiwan

Kuo-Yang Tu — National Kaohsiung First University of Science and Technology, Taiwan

Regional/Invited Session Chairs

Guido Herrmann — University of Bristol, UK

Adalberto Llarena — Universidad Nacional Autónoma de México, Mexico

Peter Kopacek — Vienna University of Technology, Austria

Norber Michael Mayer — National Chung Cheng University, Taiwan

Qixin Cao — ShangHai Jiao Tong University, China

Program Co-chairs

Ching-Chih Tsai — National Chung Hsing University, Taiwan

Chen-Chien Hsu — National Taiwan Normal University, Taiwan

Chien-Cheng Tseng — National Kaohsiung First University of Science and Technology, Taiwan

Reception Co-chairs

Chin-I Huang	National Kaohsiung First University of Science and Technology, Taiwan
Chia-Yen Chen	National University of Kaohsiung, Taiwan

Local Arrangements Co-chairs

Chao-Lieh Chen	National Kaohsiung First University of Science and Technology, Taiwan
Kuo-Ching Tseng	National Kaohsiung First University of Science and Technology, Taiwan
Pao-Lung Chen	National Kaohsiung First University of Science and Technology, Taiwan
Tung-Kuan Liu	National Kaohsiung First University of Science and Technology, Taiwan
Tom Tsai	National Kaohsiung First University of Science and Technology, Taiwan

Website Chair

Haw-Ching Yang	National Kaohsiung First University of Science and Technology, Taiwan

Conference Committee

CIRAS - International Conference on Computational Intelligence, Robotics and Autonomous Systems

General Chair

Tzuu-Hseng S. Li	National Cheng Kung University, Taiwan

Program Chair

Prahlad Vadakkepat	National University of Singapore, Singapore

ICAHRR - International Conference on Advanced Humanoid Robotics Research

General Co-chairs

Jacky Baltes	University of Manitoba, Canada
Kuo-Yang Tu	National Kaohsiung First University of Science and Technology, Taiwan

Program Chair

John Anderson	University of Manitoba, Canada

ICEER - International Conference on Education and Entertainment Robotics

General Co-chairs

Ching-Chang Wong   Tamkang University, Taiwan
Norbert Jesse   Technische Universität Dortmund, Germany

Program Chair

Chung-Hsien Kuo   National Taiwan University of Science and
           Technology, Taiwan

Program Committee

Z. Llarena, Mexico
Adbullah Al Mamun, Singapore
Nadir Ould-khessal, Canada
Chung-Hsien Kuo, Taiwan
Pramod Kumar Pisharady, Singapore
Guido Bugmann, UK
Guido Herrmann, UK
Igor Verner, Israel
Joerg Wolf, UK
Vivekananda Shanmuganathan, India
Kok Kiong Tan, Singapore
Tung-Kuan Liu, Taiwan
Yung-Tien Liu, Taiwan
Pei-Yung Hsiao, Taiwan
Bao-Long Jhang, Taiwan
Wen-Chung Chang, Taiwan
Chang-Jung Juan, Taiwan
Guo-Lan Su, Taiwan

Meng Cheng Lau, Canada
Chi-Tai Chen, Taiwan
Nils Axel Andersen, Denmark
David J. Ahlgren, USA
Gourab Sen Gupta, New Zealand
Ranjan T.N., India
Simon Parsons, USA
Tony Dodd, UK
Ulrich Rueckert, Germany
Julien Diard, France
Z.Y. Dong, Hong Kong
Jyh-Cheng Yu, Taiwan
Wen-Long Yao, Taiwan
Chih-Hung Wu, Taiwan
Chen-Chia Chuang, Taiwan
Yih-Guang Leu, Taiwan
Jin-Tsong Jeng, Taiwan

Table of Contents

A Motion Tutoring System by Using Virtual-Robot and Sensors

Tae-Jin Kim, Kyoung-Tae Lee, and Nam-Hyeok Kim

Department of Electrical Engineering and Computer Science,
KAIST, 355 Gwahak-ro, Yuseong-gu, Daejeon, Republic of Korea
{kimtj5521,kyu461,walwal4q}@kaist.ac.kr

Abstract. This paper describes how the user can determine the exactness of his motion by using proposed tutoring system based on virtual robot without any other's help. Sensors in the user's cloth measure the positions of the joints. The main PC gathers these data and shows the user's motion by using virtual robot. Tutoring system in PC compares between the database of exemplary motion and the user's motion and gives feedback to the user.

Keywords: Motion Capture, 3D Avatar Rendering, Tutoring by Virtual Robot.

1 Introduction

In the generation that smart phones containing the ability of computer are generalized, people's interest in smart environment has increased every day. Many devices tend to be miniaturized and lightened. The concept 'Wearable Computer' is no more strange today. Although they are not commercialized yet, lots of wearable computers such as OLED glasses which shows valuable data at the user's retina or a cloth which can control MP3 player and cell phone wirelessly by using Bluetooth are developed. Many researchers are trying to develop many efficient and comfortable wearable computers these days [1]–[5]. Following this tendency, this paper proposes a motion tutoring system by using virtual robot and sensors. This system is composed of motion capturing by using gyroscope and accelerometer, displaying the user's performance through avatar, comparison between virtual robot and the database containing the exemplary motion, and feedback process which shows the exactness of the user's motion in percentage scale.

Many researches related to motion capturing have been conducted. Especially, motion capturing using camera is the most widely used. This technology was used in the movie 'Avatar'. However, this technique is often vulnerable to ambiguities in the video data (e.g., occlusions, cloth deformation, and illumination changes), degeneracy in camera motion, and a lack of discernible features on a human body [6]. Also, the technique is not real time because a scene has to be taken once, translated into three dimensional, and it has to be taken again to proceed to the next scene. Furthermore, since some parts not caught by camera would be modeling as approximated value, it is not exact way to capture the user's

T.-H.S. Li et al. (Eds.): FIRA 2011, CCIS 212, pp. 1–7, 2011.

motion. These drawbacks can be solved when gyroscope and accelerometer are used simultaneously. Since the user moves sensors on his joints directly, the avatar moves exactly and in real time. An experiment was conducted in order to demonstrate walking motion of the user.

The rest of this paper is organized as follow. In Section 2, the details of the tutoring system are presented. Section 3 describes the implementation of over-all system. The experimental results are discussed in Section 4 and concluding remarks follow in Section 5.

2   Tutoring System

The basic concept of tutoring system is comparison of motion difference between virtual robot which moves as in database and avatar which makes same motion measured from the user. Both data types of databases and measured motions are transferred angle-based data. The avatar and the virtual robot are rotated by the angle-based data. Comparison process between avatar and the virtual robot is using special judgment system. The user can get feedbacks by watching the real time score which displays the exactness of the user's motion. And this process is repeated in order to tutor next motion of the user.

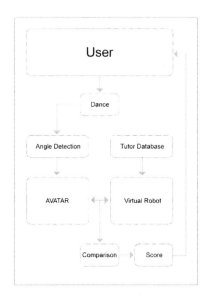

Fig. 1. Tutoring system architecture

2.1   Structure of Virtual Robot

The tutoring system is accomplished by the comparison between the user's avatar and virtual robot. For the efficiency of comparison, avatar and virtual robot are designed as same structure. Fig. 2 represents the basic structure of virtual robot.

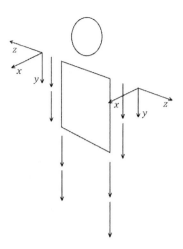

Fig. 2. Virtual robot structure

Virtual robot has eight unit vectors whose magnitude is one. Two unit vectors form one set. So the virtual robot has four sets of vectors. Each set represents left arm, right arm, left leg, and right leg. It splits into two vectors by an elbow and knee.

In Fig. 2, the axis in the right side of robot is different from the left side. The reason is that these axes are considered in real module sensors. The system assumes that the initial posture of the user is same as Fig 2. Because of this reason, the initial values of all unit vectors are (x, y, z) = (0, 1, 0).

2.2 Judgment System

Judgment system is accomplished in each four sets of vectors. The basic criterion of judgment is to measure the distance between each joint (elbow, wrist, knee, and ankle). In Fig. 3, vector v_1 represents the part of upper arm (from shoulder to elbow) and vector v_2 represents the part of lower arm (from elbow to wrist). The vectors which has prime (') represents the vectors of virtual robot. Then the judgment of internal joint (elbow and knee) can be represented as comparison vector d_1 which is defined as

$$d_1 = v_1' - v_1. \tag{1}$$

Also, the judgment of external joint (wrist and ankle) is achieved by comparison vector d_2 which is defined as

$$d_2 = (v_2' + v_1') - (v_2 + v_1). \tag{2}$$

The judgment function has three significant characteristics. The first one is that the judgment range is limited up to 90 degrees. The second characteristic is that the judgment result is graduated linearly by the distance. The last one is that the range of the result is 0 to 100. It means that if the angle difference

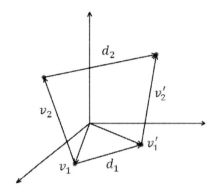

Fig. 3. Judgment method

between two unit vectors is 90 degrees, the result becomes 0. Similarly, if there is no difference, the result becomes 100.

When the angle difference is 90 degrees, the distance between those two becomes $\sqrt{2}$. In that sense, the linear judgment function which has a comparison vector d as a variable and satisfies all the characteristics is derived as

$$f(d) = \max\left\{ \frac{(\sqrt{2} - |d|)}{\sqrt{2}} \, , \, 0 \right\} \times 100\%. \tag{3}$$

Through this judgment function, the result of judgment is calculated and showed to the user in real-time.

3   Implementation

3.1   Expression of Data

Processing the received values from modules, it finally becomes to the rotated angles in three orthogonal axes in every 40 milliseconds. This processed information is formed as Euclidean coordinates (pitch, roll, and yaw). The tutoring system uses this data structure. Since there are twelve modules in the suit, the total number of the data in floating number becomes thirty six. These thirty six floating numbers are formed as one data set for 40 milliseconds.

3.2   Implementation of Rotation

The implementation of rotation is achieved by Euclidean coordinates data sets. With these data sets, the unit vectors of the avatar and virtual robot are rotated. For making data sets into rotating vector, 'Quaternion' concept is used. The reason that the system uses quaternion is to eliminate the ambiguities in rotation. With the pitch, roll, and yaw data, system makes the quaternion which has same meaning. After that, the system extracts a rotation vector from the quaternion. By multiplying the rotating vectors which are resulted from these procedures by the vectors of the avatar and the robot, the rotation can be implemented.

3.3   Sensors

Gyroscope and acceleration sensor are put on joints in order to capture the user's motion. Gyroscope measures the angular velocity and acceleration sensor measures the amount of acceleration. So, the amount of angle rotated can be measured by integrating angular velocity data from gyroscope. Also, a slope can be measured by using acceleration data. With these data, position data is measured to make an avatar of the user.

3.4   Filter

Even though the position data can be measured by just using gyroscope or acceleration sensor alone, there are also disadvantages. Gyroscope is weak at some disturbances such as temperature variation. Also, we can't measure the slope of joints correctly by using acceleration sensor alone when the user moves fast because acceleration of joints is added. In order to remedy these disadvantages, simple filter can be used. Disturbances of gyroscope can be lessened by passing through a high pass filter because they vary slowly. Also, pure slope data can be obtained by passing the acceleration data through a low pass filter. After combining these filtered data, more accurate position information is measured.

3.5   Overall System Flow

There are 12 modules on the user's body. Each module is placed at wrist, elbow, shoulder, ankle, knee, and waist of the user. Three modules form one module set so there are four module sets. Each module samples the angle data of joints. Root module gathers these data by using UART module and transmits them to PC wirelessly. Sampling frequency is 25Hz. Zigbee module is used in order to communicate stably.

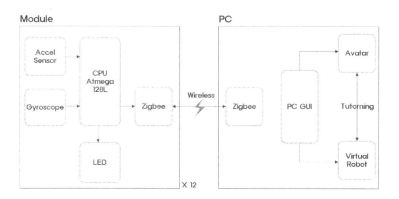

Fig. 4. Overall system flow

3.6   Miscellaneous Things

Each module has 12 LEDs in order to maximize visual effect of dancing. Series connected two 1.5V AA rechargeable batteries provide power to each module set. Since every chip is driven by 3.3V, we used MAX1676 regulator.

4   Experiments

To demonstrate the proposed system, a user performed a walking motion with sensor suit and checked avatar through the virtual robot tutoring system. In the software development environment, Windows Form (CLI) was used in order to make Graphic User Interface (GUI). CLI is provided by .Net Framework 4.0 which is one of Windows Application Interface (WinAPI) development systems. To draw an avatar, OpenGL was used. Development tool was Visual C++ 10.0. Data from sensor module were transmitted to main program through Serial Port. Main program gathered these data every 40ms, processed them, and rotated the joints of the avatar. Simultaneously, it got the data of exemplary motion from database and rotated the joints of virtual robot. Finally, it compared them and gave feedback to user.

In the experiment, the user raised right arm and left knee as he is walking. The tutoring system compared avatar with virtual robot which had database of walking motion. In the first step, eight units' score were 100 percent. After several seconds, virtual robot raised left arm and right knee. Simultaneously, virtual robot put right arm and left knee down. However, the user took the first motion continuously. As shown in the Fig. 5, score dropped because the motion of avatar did not correspond with virtual robot data. Also after several seconds, virtual robot reversed the motion and avatar corresponded with virtual robot data. Eight units' score returned 100 percent in real time.

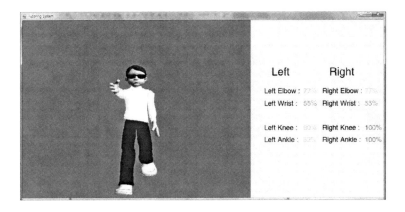

Fig. 5. Snap shot of experiment result

5   Conclusion

A motion tutoring system by using virtual-robot and sensors was proposed in this paper. With this system, the user could check his exactness of motion alone. When the user wore proposed suit which contained gyroscopes and accelerometers and moved, they PC checked the exactness of the user's motion and gave feedback. When the experiment which checked the walking motion of the user was conducted, proposed motion tutoring system worked successfully. So, the user could improve his imperfections of motion by using it. Now, the sensor suit could detect the rotation of arm and leg, not that of wrist. However, if several modules are added and elaborate motion capturing technique is developed, it could be used at rehabilitation or some kinds of sports which require accurate motion.

References

1. Semwal, S.K., Hightower, R., Stansfield, S.: Mapping Algorithms for Real-Time Control of an Avatar Using Eight Sensors. Presence 7(1), 1–21 (1998)
2. Lee, S.W., Mase, K.: Activity and Location Recognition Using Wearable Sensors. IEEE Pervasive Computing, 24–32 (2002)
3. Pentland, A.: Looking at People: Sensing for Ubiquitous and Wearable Computing. IEEE Trans. on Pattern Analysis and Machine Intelligence, 107–119 (2000)
4. Starner, T., Mann, S., Rhodes, B., Levine, J., Healey, J., Kirsch, D., Picard, R., Pentland, A.: Augmented Reality Through Wearable Computing. Presence: Teleoper. Virtual Environ. (1997)
5. Kern, N., Schiele, B., Schmidt, A.: Multi-sensor Activity Context Detection for Wearable Computing. In: Aarts, E., Collier, R.W., van Loenen, E., de Ruyter, B. (eds.) EUSAI 2003. LNCS, vol. 2875, pp. 220–232. Springer, Heidelberg (2003)
6. Wei, X., Chai, J.: VideoMocap: Modeling Physically Realistic Human Motion from Monocular Video Sequences. In: ACM TOG (Proc. SIGGRAPH 2010) (2010)
7. Using Quaternions to Represent Rotation, http://gpwiki.org/index.php/OpenGL:Tutorials:Using_Quaternions_to_represent_rotation
8. Creating an OpenGL View on a Windows Form, http://www.codeproject.com/KB/miscctrl/OpenGLViewWinForms.aspx

An Adaptive Component Model for Autonomous Mobile Robots[*]

Jinhui Zhu[1], Huaqing Min[2], Yingju Liang[1],
Mingjie Liang[1], Chang'an Yi[1], Mei Zhang[3], and Junping Li[4]

[1] School of Computer Science and Engineering,
[2] School of Software Engineering,
[3] School of Automation Science and Engineering,
South China University of Technology,
Guangzhou, China
[4] Shenzhen Polytechnic,
Shenzhen, China
{csjhzhu,hqmin}@scut.edu.cn
{yju.liang,mjie.liang,yi.changan}@mail.scut.edu.cn,
zhangmei@scut.edu.cn,
ljp@szpt.edu.cn

Abstract. In order to achieve flexibility and reusability in mobile robot development, we present a component-based robotic framework with its adaptive component model, MiniROS. It provides data, signal and service port as inter-component interfaces so that developers can design complex robot behaviors with data-based, control-based or service-based paradigm. In addition, a strategy is introduced to allow member components of a robot to automatically adjust the structure and parameters by themselves according to the environment and tasks. Finally we show an example application using our component-based model.

Keywords: robotic framework, adaptive component model, autonomous mobile robots.

1 Introduction

Nowadays robot applications are facing difficulties from incompatible enviroment and complex tasks, which brings about increasing complexity to robot softwares. Therefore, a standard software framework is required to facilitate the robot software development. Middleware technology is a reasonable solution. There are a few advantages of middleware, such as shortening of the development cycle and improving the software quility. For this reason, robot middleware technology is becomming a research hotspot for autonomous mobile robots.

---

[*] This research is supported by the Fundamental Research Funds for the Central Universities (No. 2009ZM0297) , National Natural Science Foundation(61005061) and Science and Technology Planning Project of Guangdong Province, China (2010B010600016).

T.-H.S. Li et al. (Eds.): FIRA 2011, CCIS 212, pp. 8–16, 2011.

Several robot middleware platforms have been proposed in recent years, for example Player/Stage[1], Miro[2][3], Orca[4], OROCOS[5], OpenRTM-aist[6], ROS[7] and so on. Mohamed and Al-Jaroodi summerized some typical platforms[8]. Most of them are designed as a component-based system, defined their own component model, and provided facilities of resource management and communication. Thanks to these software engineering approachs, reusablity, flexiblity and scalability of robot software have been significantly improved.

In pratice, robot behavior is designed for specific environment and tasks. As a result, robot software need to adaptively change its software architecture and adjust its parameters. At present, these adaptive strategies is usually hard-coded to robot software by its developers. There has been lack of support on the level of robot software platform.

With this paper, we present a new adaptive component model in MiniROS framework for autonomous mobile robots software. The rest of the paper is organized in four sections. Section 2 is on previous work done in the robotic community. In the third section, we present the architecture of MiniROS. The adaptive component model are presented in the fourth section. In section 5, we illustrate how to build robot applications with MiniROS as an example.

2   Related Work

In recent years, software technology has led to the improvement of robotic technology. However, more and more complex and advanced task requirement results in sophisticated robot software development. In robotic community, researchers presented several robot software frameworks in order to fulfill the requirement, which will be introduced as follows:

The main goal of the **OROCOS[5]** project is to develop a general purpose modular framework for robot and machine control. OROCOS uses a C++ class TaskContext to define its component, in which five types of interfaces can be defined: Event, Attribute Properties, Commands, Methods and Data-Flow Ports.   Inside the component, state machine is the main execution model. However, the communication pattern in OROCOS is too complicated, also, the connections between components must be set up by hand-written code.

OpenRTM-aist[6] is open source implementation of the RT-Middleware specification and is developed by AIST, Japan. Component within OpenRTM can interact either through a client/service pattern for service ports, or publish/subscribe pattern for data ports, and it also provide configuration interface to modify the parameters of core logic from outside the component. The component execution model is also based on state machine.

The **Robot Operating System (ROS)[7]** ROS was designed to meet a specific set of challenges encountered when developing large-scale service robots at Stanford University and the Personal Robots Program at Willow Garage. ROS provides a structured communications layer above the host operating systems of a heterogenous compute cluster. In ROS all the infomation exchange among nodes is performed through messages. Nodes can join and leave the system dynamically. ROS is a suitable tool to integrate other existing robot softwares and libraries, nevertheless, it is not a real component-oriented software, since it does not define any specific computation unit model inside so far.

We have present a component-based software architecture for multi-mobile robots, **CSAMR**[9] before. In the model of CSAMR, different components of sensor, actuator, decision and etc, are integrated into an agent for a special function robot. Several agents complete a task by collaboration. Components and Composite components are self-described in XML file. Composite component is the container and manager of components. However, the interfaces of CSAMR are defined as service, which is not sufficient for express complex robot behaviors, and also, it does not provide adaptive mechanism in its component model.

3   MiniROS Architecture

In order to support the development of complex and variable robot software, our framework must fulfill the following requirment:

* Seperation of algorithm interface and implementation;
* Composite component support;
* Multi-robot system: component can join or leave system dynamically, and error-detecting and error-recovering ability are required ;
* Flexible architecture design: easy to organize components with hierarchical, reactive or hybrid deliberative/reactive paradigm in robotics ;
* Adaptive evolution capability: robot should change the structure or parameters of itself according to the enviroment and task.

The majority of the frameworks mentioned in the previous section, meet some of the design objectives. However, none of them fulfills the full list of the design goals, expecially in multi-robot system and adaptive evolution support.

MiniROS is a component-based robot framework, which aims to privode support for reusable robot software, with an adaptive component model. The architecture of MiniROS is shown in Fig. 1.

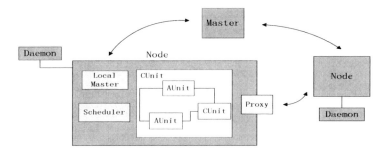

Fig. 1. MiniROS Architecture

A component in MiniROS architecture is called a **Unit**, which is the basic reusable element. Unit can be a composition of other units, which allows developers to design system with hierachical abstraction. **Node** is the rumtime environment and life cycle manager of Units. It corresponds to a process of operating system. The structure of Node is shown in Fig. 1, in which scheduler is responsible for running units, LocalMaster helps Units to interact with each other, while Proxy makes possible for units to communicate through the network. As the manager of Nodes, **Master** assists in the discoveries and connections among Nodes, starts or stops the specific node in robot application system, and monitors the health of each Node. **Daemon** is the program running in the background of computer. It monitors and reports the status of Nodes, and helps the Master to manage the life cycle of Nodes.

4 Component Model

4.1 Component Interface

Component interface is the channel to exchange data with outside the component. Three types of interface are introduced to MiniROS (see the Fig. 2): **Data ports** includes Input and Output ; **Signal ports** have two types, SignalIn for receiving the notification and SignalOut for publishing event; **Service ports** are for service-based, in which, Caller is used to call external service, while Callee is the port providing service to external components.

Those three types of interface cover most of the communication patterns, such as Event, Command, Method and Data flow.

4.2 Atomic Unit

Atomic Unit is basic functional component which is compiled from source code by developers. An atomic unit has three parts: interfaces, properties, and atomic behavior. Interfaces are made up of the three types component interface mentioned above. Properties are the configuration of atomic units, Atomic behavior is the fundamental element for function implementation and scheduling. Three types of atomic behaviors are defined in MiniROS: Normal Behavior, Service Behavior and View Behavior:

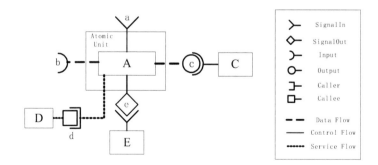

Fig. 2. The Atomic Unit Model. A, E are normal behaviors, D is service behavior, and C is view behavior. b->A->c->C is the data flow, a->A->c->E is the control flow, D->d->A is the service flow.

* **Normal Behavior:** the semantics in a normal behavior is that, when receiving the signals of all SignalIn ports, read the data from Input ports, process them with the help of external services through Caller ports, finally output the result to Output ports, and send signals through its SignalOut ports. (see A in Fig. 2)
* **Service Behavior:** provides a service to outside the component. It must bind to a Callee port. (see D in Fig. 2)
* **View Behavior:** A window is contained in it to deal with user interaction. It automatically translates the data from Input ports to messages and send them to window, then the window process and visualize the message; The user operations can be translated back to MiniROS data and send out through Output ports automatically.

It should be mentioned that SignalIn ports can be connected with Output ports, hence the data from Output ports can be seen as a signal to other components.

With these three types of behaviors, developers are able to implement applications with data-flow based, control-flow based and service based design paradigm.

4.3 Composite Unit

A Composite Unit consists of interfaces, properties, composite behaviors and strategy. The interfaces and properties are referenced to its internal units. Composite Behavior is a set of atomic units and composite units, which represents a function of a composite unit.

In MiniROS, Strategy is a powerful mean for robot software adaptive evolution. Strategy is comprised of several rules with the form of "condition-action". The grammar of strategy is shown following:

The EBNF grammar of strategy

```
Strategies = {StrategyRule}
StrategyRule =
"When" , ConditionPredicate ,
"Do" , "{" ,
    Action, {"," , Action} ,
"}"
```

The condition predicates in MiniROS strategy mechanism includes:
* Predicates about the behavior of atomic units, such as computation time limit, behavior finish, behavior error and so on;
* Interface predicates, for example predicates about the data in Input ports, or predicate whether the SignalIn port is triggered;
* Predicates describing the unit state, for example unit activated, unit stopped.

The actions supported by MiniROS are list following:

* Controlling life cycle of units, including load, activate, deactivate, unload;
* Management of composite behavior, including load, activate, deactivate, unload, add unit and delete unit;
* Properties change of internal units;
* Connection or disconnection of internal units;

Thanks to the MiniROS strategy mechanism, composite unit can adjust its internal behavior and unit structure, or change the configuration parameters, depending on the change of environment and task.

Since composite unit is essentially a configuration of existing units, it doesn't have any binary code. Therefore, changing of composite unit is quite simple, and this bring great flexibility to MiniROS.

5 Application of MiniROS

Machine vision is a common problem in robotics. It aims to extract the information and features of objects that robots are interested in, through the method of image processing and analysis.

In general, vision system has three key components: image capturing, image processing and visualization. It is expected that these three components are organized in a flexible way, which means that the architecture of vision system can be changed conveniently. Another requirement is that vision system can choose the process algorithm itself, dynamically and automatically, based on the environment (for example the device speed). The MiniROS framwork facilitates the building of such a system. Fig. 3 is an example of our vision system, built with MiniROS, which is running on the soccer robot.

Fig. 3. Soccer robot and vision image. Camera is equipped on the top, and the vision system runs in the computer installed in the body of robot. The vision system analyzes the image, and find out the obstacles around the robot to avoid collision with other objects.

5.1 Building the Atomic Units

Image capturing and image processing are done by Camera unit and Vision Processor unit (shown in Fig.4), respectively. Camera unit is the abstraction of hardware, that calls the specific camera driver to capture the image and output to the port 'pic'. Vision Processor is a unit that do the realtime processing with the image, then output the object infomation that it recognize. In order to get the camera infomation before capturing, Vision Processor requires an initialization behavior, called 'init'.

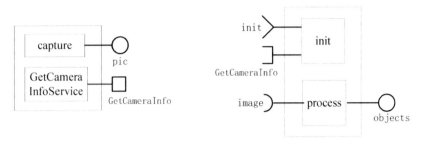

Fig. 4. Camera unit and Vision Processor unit

The visualization of the vision system is handled by units with view behavior. Image Viewer is to show the image from the camera, while Object Viewer displays the object information.

After design of the atomic units, MiniROS will generate the code framework. Developers need to implement the specific functions, and compile the atomic unit. The code generated from the Vision Processor is list following.

Code Generation of VisionProcessor. It is translated to a sub class of UserAtomUnit, and it is required to implement four call-back functions: onLoad, onUnload, onActivate, and onDeactivate, which are called in relevant phase. The declaration code of interface and property is generated in onLoad function. The other two function, init and process, implement the behaviors of unit, and the correspond ports is translated to the function parameters.

```
class VisionProcessor:public UserAtomUnit {
public:
    Property<int> name;
    VisionProcessor ();
    ~VisionProcessor ();
    void onLoad();
    void onUnload();
    void onActivate();
    void onDeactivate();
    void init(Caller& GetCameraInfo);   // init behavior
    void process(const Image* image,
                        Image* classified_image,
                        Objects* objects;
                ) ;   //vision process behavior
};

void onLoad() {
    Publish<Objects>('objects') ;
    Subscribe<Void,CameraInfo>('GetCameraInfo') ;
    DeclarePropety(name, 'name') ;
    //user inintialization code
}
```

5.2 Construction of Composite Unit

In Vision Composite Unit, we use Vision Behavior to wrap the main functions of vision system, including camera and vision processor, as shown in the left side of Fig. 5. The left-top corner of vision behavior is its initailization port, which signals when the behavior initializes, and then cause the execution of 'init' behavior in vision processor.

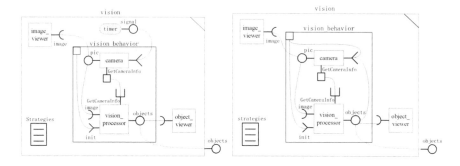

Fig. 5. the Vision Composite Unit. In the left figure, the timer triggers the camera unit to capture the image in this composite unit periodically, while in the right figure is a loop architecture. Image data is transffered to vision processor, and finally vision processor do the calculation and output the object information it recognize. The 'objects' port of composite unit is referenced to the port of vision processor, so as to privide the object information outside the vision composite unit.

Flexibile architecture capability is one of the key characteristics of MiniROS. For the vision system, we can simply change the timer-driven architecture to the loop architecture. As shown in the right side of Fig. 5, the initialization of vision behavior drives the first execution of camera, and after that the data produced by vision processor unit, as a signal, activates the next execution of camera.

With the strategy mechanism of composite unit, MiniROS has powerful ability to support dynamic evolution. When the classic algorithm in vision processor cannot satisfy the deadline restriction, a rule of changing unit will be activated, to stop and unload the existing vision process unit, load a faster unit, and continue to run.

A Strategy Rule of Changing Vision Processor

```
When vision_processor.process.deadline_unsatisfied(100)
Do{
        DeactiveUnit vision_processor
        UnloadUnit vision_processor
        LoadUnit fast_vision_processor into
vision_behavior
        ActiveUnit fast_vision_processor
}
```

6 Conclusions

In this paper, we propose a new adaptive component model for autonomous mobile robots for MiniROS framework. It is designed for fast development of flexible and reusable robot software. Component is called Unit in this model. The component interface of unit, including data port, signal port and service port, provides adequate patterns to meet the communication requirments. In atomic unit, developers can use normal, service and view behavior to design component with different paradigm. In composite unit, strategy is introduced to adaptively control the adjustment of structure and parameters, according to the environment and tasks given to a robot. In general, MiniROS component model provides an easier way to build flexible and adaptive robotic systems.

References

1. Kranz, M., et al.: A Player/Stage System for Context-Aware Intelligent Environments. In: Proceedings of UbiSys 2006, System Support for Ubiquitous Computing Workshop, Orange County California (2006)
2. Enderle, S., Utz, H., Sablatng, S., Simon, S., Kraetzschmar, G., Palm, G.: Miro: Middleware for autonomous mobile robots. In: IFAC Conference on Telematics Applications in Automation and Robotics (2001)
3. Utz, H., Sablatng, S., Enderle, S., Kraetzschmar, G.: Miro – Middleware for Mobile Robot Applications. IEEE Transactions on Robotics and Automation 18(4), 493–497 (2002)
4. Makarenko, A., Brooks, A., Kaupp, T.: Orca: Components for Robotics. In: International Conference on Intelligent Robots and Systems (IROS), pp. 163–168 (2006)
5. Bruyninckx, H.: Open robot control software: the OROCOS project. In: IEEE International Conference on Robotics and Automation (2001)
6. Ando, N., Suehiro, T., Kitagaki, K., Kotoku, T., Yoon, W.-K.: RT-Middleware: Distributed Component Middleware for RT (Robot Technology). In: 2005 IEEE/RSJ International Conference on Intelligent Robots and Systems (IROS 2005), Edmonton, Canada, pp. 3555–3560 (2005)
7. Quigley, M., Conley, K., Gerkey, B.P., Faust, J., Foote, T., Leibs, J., Wheeler, R., Ng, A.Y.: ROS: an open-source Robot Operating System. In: ICRA Workshop on Open Source Software (2009)
8. Mohamed, N., Al-Jaroodi, J., Jawhar, I.: Middleware for Robotics: A Survey. In: IEEE Conference on Robotics, Automation and Mechatronics (2008)
9. Zhu, J.H., Min, H.Q., Feng, F., et al.: Design of component-based software architecture for multi-mobile robots. Journal of South China University of Technology (Natural Science) 36(1), 13–17 (2008)

Toward Safe Human Robot Interaction: Integration of Compliance Control, an Anthropomorphic Hand and Verbal Communication

Said Ghani Khan[1], Alexander Lenz[1], Guido Herrmann[2],
Tony Pipe[1], and Chris Melhuish[1]

[1] Bristol Robotics Laboratory, University of Bristol and University of the West of England, Bristol, UK
[2] Department of Mechanical Engineering and Bristol Robotics Laboratory, University of Bristol, UK

Abstract. In this paper an integrated system for human robot interaction is presented. It is demonstrated that safety features in human robot interaction can be engineered by combining a robotic arm, equipped with a compliant controller, an anthropomorphic robot hand and a spoken language communication system. A simplified human-robot interaction scenario, based on a typical care robot situation, is exploited to show that safety can be enhanced by the monitoring of torques and motor currents to establish contact with the environment. Furthermore, spoken language is utilised to resolve potentially dangerous contact situations.

1 Introduction

Cooperative Human-human interaction features a range of communication methods which, due to their partial redundancy, help to make interaction in diverse, noisy and chaotic environments possible. With the emergence of robotic carers or robotic assistance, which will operate with or close to humans, it is desirable that such robots are equipped with similar communication and sensing capabilities as humans. This will not only allow the human to interact 'naturally' with the robot, but also contributes, due to the communication and sensing redundancy, to safety during interaction.

In this paper, we present an integrated system which has been successfully employed in a safe object passing task between a human and a robot. The system consists of (i) an anthropomorphic hand, comprising capabilities like grasping, pointing, releasing and 'sensitive' hand-over (see section 3); (ii) a humanoid robotic arm fitted with a model reference adaptive compliance controller with contact detection, running on a dSPACE system (see section 5); (iii) a bi-directional spoken language interface to issue commands, report states and request instructions (see section 4). We show how those subsystems, all running on different computational platforms, can be integrated, and demonstrate the system's functionality with an interaction scenario that utilises all (built in) modalities.

T.-H.S. Li et al. (Eds.): FIRA 2011, CCIS 212, pp. 17–24, 2011.
© Springer-Verlag Berlin Heidelberg 2011

The paper is organised as follows. In the next section we introduce the interaction scenario. Afterwards, the robotic subsystems and their integration are outlined. The paper then concludes with the illustration of the results and a brief discussion.

2   Interaction Scenario

For the experimental work presented here, we have designed a simplified human-robot interaction scenario. Although this only demonstrates a small aspect of human-robot interaction that would be desirable in future care robots, it is representative for situations where a robot assists a physically less-able and possibly bedridden human.

Imagine the following care situation: A bedridden human should be able to direct the robot towards a set of locations and instruct the robot perform a certain task at a given location. During the repositioning of the robot, collisions can occur and the robot would need to be able to sense those, reduce the impact force to a minimum, report the collision to the human and ask for advice on how to proceed. Handling task that the robot performs should also be sensitive to impact. The hand-over of objects is of particular importance and the robot would be required to only release objects when the human holds them firmly.

Our simplified lab scenario reflects this general purpose example. A human is placed at a chair and the BERT2[1] robotic arm is mounted on a stand. A paper bin is placed out of reach of the human but within reach of the robot (see figure 1). The robot arm can extend its hand to the human. In our scenario, both of those locations (i.e. 'close to human' and 'above bin') are pre-programmed, but could easily be dynamically set or acquired by a higher level perception and reasoning system, as shown before by the authors[2]. The human can now verbally instruct the robot to move to him/her or to the bin. Furthermore, it is possible to instruct the robot to take an object, release an object or hand-over an object. The difference between release and handover is that in the former the robot hand simply opens, whereas in the latter the hand opens only when the

Fig. 1. BERT2 hands a cup to a human experimenter

hand's sensing system perceives firm human contact with the handled object. During all those interactions, the robot also verbally announces its next action and position, which raises the human's awareness and therefore also enhances safety.

During the arm motion, the adaptive controller dynamically estimates the torques in the arm, which are subject to joint angles, velocities and accelerations. If the measured torques exceeds the torque estimate, a collision is assumed. The robotic system verbalises this to the human and waits for further instructions. The human can then either instruct the robot to stay were it is or alternatively return to the last established position before a motion command was issued. This introduces an important safety feature in addition to the active compliance of the robot arm.

3   Grasping Controller for an Anthropomorphic Hand

For the grasping of objects, a single anthropomorphic hand[1] module (Fig. 2) has been integrated into the arm control infrastructure. The hand has nine degrees of freedom (DOF) comprising gross movement for all 5 fingers, "trigger" action for index and middle finger, opposing of the thumb, and finger spread. Each DOF is actuated by a geared brushless DC motor and the lowest level motor control is established via EPOS[2] units, which are configured as velocity controllers in a 1kHz loop. All motors and gearing are fully integrated into the hand and (in contrast to many cable driven hands) no mechanical parts require integration into the robot arm. To facilitate simple integration into a range of robotic platforms, and to allow higher-level hand movement commands (e.g. grasp(.),open(.),point(.),handOver(.),) to be executed from a host system, a independent control and drive system has been developed. This embedded microcontroller platform is based on the dsPIC30F6014A digital signal controller from Microchip[TM]. The board is equipped with one CAN interface for communication with the individual EPOS drivers for each DOF. The embedded software takes care of homing, calibration of the individual joints, and issues the velocity demands based on the current position and the higher level demand. Access from and to higher level systems can be established via the on-board RS232, SPI, a second CAN, or Ethernet interfaces. In our application, the dSPACE system communicates with the hand controller via the second CAN interface. Additionally, this subsystem features 16 analogue channels which will be utilised with the employment of touch sensor on fingers and palm in the future.

When a high-level grasping command (grasp(.)) is issued, the hand controller starts to close the fingers and monitors the motor currents. If they exceed (individually configurable) current thresholds, the actual finger position is stored

---

[1] The mechanical design and manufacturing for the BERT II arm and hand has been conducted by Elumotion (www.elumotion.com); the embedded driver board and the software has been developed by the authors.

[2] Maxon[TM] manufactures a range of EPOS controllers. We use low power range for each DOF of the hand, the EPOS 24/1.

Fig. 2. The BERT2 anthropomorphic hand (right) during a object hand-over procedure

and set as a new reference position. Since this is done for each finger individually during concurrent motion, the hand is able to grasp objects of varying size and irregular shapes reliably. This not only illustrates the versatility of the developed software but also emphasis the utility of the anthropomorphic design.

The handover command (`handover(.)`) uses the motor currents to dynamically release the object, and no touch sensor are required for this. When the handover command is issued, the controller calculates the norm of the motor currents (averaged over a 100ms time windows). This value is stored and then cyclicly ($T_{cycle} = 10ms$) compared to the actual norm of the motor currents. If this measure exceeds a pre-set deviation threshold, an open command (`open(.)`) is issued and the hand opens. For instance, in practice, all that is required for a human interactor, to slightly pull (or otherwise manipulate) the object, the robot's hand is grasping. This leads to a variation in the motor currents which the mechanisms described above detects. Consequently, the hand controller instigates the release of the object. This simple mechanism is very robust and by its nature self-calibrating.

4   Spoken Language Interface and High Level Control

In order to enable spoken language interaction we have integrated the CSLU Toolkit [3] Rapid Application Development (RAD) into the arm control infrastructure. RAD allows the use of the TCL scripting language which permits binding with a YARP[4] 'backbone' that then establishes the bi-directional communication with dSPACE. RAD and YARP are installed on a Windows^{TM} XP PC and communication to dSPACE is established via a $115.2kbit/s$ RS232 connection. RAD employs the Festival speech synthesis system and recognition is based on Sphinx-II. The construction of speech dialogues is supported via a state-based graphical programming environment. Here, it is also important to note that the RAD system requires no speaker specific recognition training which makes the interaction dialogues generic in terms of human users. The YARP interface in the speech dialogues allows us to generate motor commands to the arm, based on verbal human-robot interaction. In turn, the speech dialogues are dynamically adapted, based on the arm state feedback reported by the dSPACE. Hence, the highest level of control, i.e. the finite state machine the human verbally communicates with, has been implemented using RAD, from which all lower level motor commands are issued.

5   Model Reference Adaptive Compliance Control for BERT2 Arm

The BERT2 arm is controlled via a model reference adaptive controller (MRAC) [5] and [6]. It is a dynamic model free controller based on the work of Colbaugh [7, 8, 9]. The reference model is a compliant, second order mass-spring damper system, the stiffness and damping terms can be selected to achieve several different compliance characteristics. Our current arm controller uses 4 joints, namely shoulder flexion, humeral rotation, elbow and wrist flexion. This adaptive controller is used because of the unknown dynamics of the arm as well as the nature of the application (object passing) which cannot rely on a fixed dynamic model. The use of adaptive controllers could lead to actuator saturation, hence an anti-windup (AW) compensator is used to address this issue. Our controller operates in Cartesian space (x, y and z) which only requires 3DOF. The redundant 4th DOF is therefore utilised to generate human like motion in the arm by employing an effort minimizing posture controller based on the work of [10, 11] (see also [12]). The overall control scheme for the BERT2 arm is shown in Figure 3.

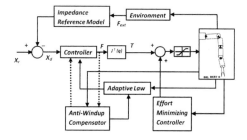

Fig. 3. MRAC based Adaptive compliance scheme with Anti-windup compensator and posture controller [5, 6, 7, 8, 9]

6   Results and Discussion

Figure 4 shows the verbal communication, arm motions and recorded forces during the interaction. BERT2 asks the human what the next action should be, reports its current position and the next action. The plots arranged in a synchronised way, illustrating the timing between verbal communication[3], the issuing of the motor commands and the actual motor action. The scenario illustrates how we interacted with the robot and how objects were grasped, released and transferred between locations (see also Fig. 1). Figure 5 illustrates a collision event. For clarity only the force acting in the z-direction is shown in 5 (c). The robot reports the detected collisions verbally to the human, while the compliant controller reduces damages during the collision. The human then directs the robot

---

[3] Abbreviations of robot utterances used in the plots: WM?="Where should I move?"; WD?="What should I do?"; MC!="I have made contact, what should I do?"

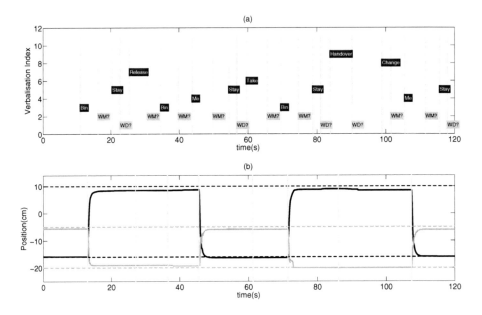

Fig. 4. Illustration of interaction scenario: (a) shows the verbal utterances of the human (white on black) and the robot (black on grey); (b) shows the robot arm motion as a consequence of the verbal communication in Cartesian coordinates. The black line represents the x-direction and the grey line the z-direction. The dotted lines represent the target positions. The y-direction was omitted for clarity. The grey vertical lines in (a) and (b) represent the times when verbal utterances and motor commands were issued.

to return to the last position (at $t \approx 38s$) or to stay where it is (at $t \approx 55s$), since in the latter case the actual position was very close to the target.

7   Conclusion

In this paper, we have successfully demonstrated how divers subsystems can be integrated to allow for safer human robot interaction. It is very clear that future robotic systems operating amongst humans will require advanced sensing systems (like artificial skin and high resolution finger-tip sensor) in order to successfully and safely assist humans. We have shown that even without those sensors, safety and dexterity is possible by exploiting information that can be obtained from motor currents (in case of the hand) or joint torque sensors (in case of the arm). This is not to say that more advanced sensors are not required. On the contrary, by combining advanced skin sensors with the already built-in sensing structure, future systems can be engineered for robustness, mutual enhancement and graceful degradation in the event of sub-system failure.

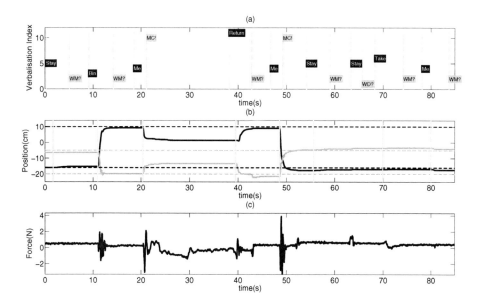

Fig. 5. Illustration of interaction scenario: (a) shows the verbal utterances of the human (white on black) and the robot (black on grey); (b) shows the robot arm motion as a consequence of the verbal communication in Cartesian coordinates. The black line represents the x-direction and the grey line the z-direction. The dotted lines represent the target positions and the y-direction was omitted for clarity; (c) shows the force acting on the robot in z-direction during collisions. The two collision events (at $t \approx 22s$ and $t \approx 50s$) are resolved, based on the human instruction, by returning to the last position or staying at the current position respectively. The grey vertical lines in (a) and (b) represent the times when verbal utterances and motor commands were issued.

Acknowledgements. The authors gratefully acknowledge the funding provide by the European Commission under the Robotics and Cognitive Systems, ICT Project CHRIS (FP7-215805). Furthermore, we would like to thank our collaborators on the CHRIS project for their insights, useful comments and challenging questions.

References

[1] Lenz, A., Skachek, S., Hamann, K., Steinwender, J., Pipe, A., Melhuish, C.: The BERT2 infrastructure: An integrated system for the study of human-robot interaction. In: 2010 10th IEEE-RAS International Conference on Humanoid Robots (Humanoids), pp. 346–351 (2010)

[2] Lallée, S., Lemaignan, S., Lenz, A., Melhuish, C., Natale, L., Skachek, S., van Der Zant, T., Warneken, F., Dominey, P.F.: Towards a platform-independent cooperative human-robot interaction system: I. perception. In: IROS IEEE/RSJ International Conference on Intelligent Robots and Systems (October 2010)

[3] Sutton, S., Cole, R., Villiers, J.D., Schalkwyk, J., Vermeulen, P., Macon, M., Yan, Y., Kaiser, E., Rundle, B., Shobaki, K., Hosom, P., Kain, A., Johan, Wouters, J., Massaro, D., Cohen, M.: Universal speech tools: The cslu toolkit. In: Proceedings of The International Conference on Spoken Language Processing (ICSLP), vol. 7, pp. 3221–3224 (1998),
http://citeseerx.ist.psu.edu/viewdoc/summary?doi=10.1.1.40.7220

[4] Fitzpatrick, P., Metta, G., Natale, L.: Towards long-lived robot genes. Robot. Auton. Syst. 56(1), 29–45 (2008),
http://portal.acm.org/citation.cfm?id=1327705

[5] Khan, S., Herrmann, G., Pipe, T., Melhuish, C., Spiers, A.: Safe adaptive compliance control of a humanoid robotic arm with anti-windup compensation and posture control. International Journal of Social Robotics 2, 305–319 (2010), 10.1007/s12369-010-0058-7, http://dx.doi.org/10.1007/s12369-010-0058-7

[6] Khan, S., Herrmann, G., Pipe, T., Melhuish, C.: Adaptive multi-dimensional compliance control of a humanoid robotic arm with anti-windup compensation. In: 2010 IEEE/RSJ International Conference on Intelligent Robots and Systems (IROS), pp. 2218–2223 (2010)

[7] Colbaugh, R., Seraji, H., Glass, K.: Adaptive compliant motion control for dextrous manipulators. The International Journal of Robotic Research 14(3), 270–280 (1995)

[8] Colbaugh, R., Glass, K., Wedeward, K.: Adaptive compliance control of electrically-driven manipulators. In: Proceedings of the 35th Conference on Decision and Control, Kobe, Japan, December 1996, pp. 394–399 (1996)

[9] Colbaugh, R., Wedeward, K., Glass, K., Seraji, H.: New results on adaptive compliant motion control for dextrous manipulators. The International Journal of Robotic and Automation 11(1) (1996)

[10] Spiers, A., Herrmann, G., Melhuish, C.: Implementing 'discomfort' in operational space: Practical application of a human motion inspired robot controller. In: TAROS Conference: Towards Autonomous Robotic Systems (August 2009)

[11] Spiers, A., Herrmann, G., Melhuish, C., Pipe, T., Lenz, A.: Robotic Implementation of Realistic Reaching Motion using a Sliding Mode/Operational Space Controller. In: Edwards, S.H., Kulczycki, G. (eds.) ICSR 2009. LNCS, vol. 5791, pp. 230–238. Springer, Heidelberg (2009)

[12] De Sapio, V., Khatib, O., Delp, S.: Simulating the task level control of human motion: a methodology and framework for implementation. The Visual Computer 21(5), 289–302 (2005)

Gait Planning of Humanoid Robots Walking on Stairs

Bi Sheng[1], Min Huaqing[2], Zhuang Zhongjie[1], Xia tuo[2], Mo Huaxi[2], Chen jian[3],
Xu Jincheng[2], Zhou Yanping[2], Li Shaojun[2], Luo Ronghua[1], and Dong Min[1]

[1] School of Computer Science and Engineering
[2] School of Software Engineering
South China University of Technology,
GuangZhou, China
[3] Shenzhen Polytechnic,
Shenzhen China
{picy,hqmin}@scut.edu.cn

Abstract. Aiming at the gain planning of SCUT-I humanoid robot, a stair motion pattern, controlled by step lengths, step period and other walking parameters, is designed. And stair restrict condition is described. According to the parameters and restrict equations, hip and ankle trajectories are planned based on IPM (inverted pendulum model). Finally, according to the model parameters of SCUT-I robot, the robot model is built and simulated by matlab6.5, which verifies the effectiveness and stability of this stairs walking gait generation method.

Keywords: Humanoid robot, Gait planning, Stairs, Parameterization.

1 Introduction

Humanoid robots are much more flexible in comparison with wheeled robots. But they are not very stable frequently during locomotion. Therefore, good gait planning are important for ensuring the stability of biped walking. There have been many studies on level walking. Only a few researchers studied the walking control in the restricted situation of circumstance, such as stairs [1].

Chenglong Fu et al. [1] generated a walking trajectory by using sensory feedback controller. Chan-Soo Park et al. [2] designed the COG (center of gravity) trajectory which is generated by the VHIPM (virtual height inverted pendulum mode) for the horizontal motion and by a 6th order polynomial for the vertical motion. Figliolini et al. [3] developed the biped robot EP-WAR3, which can walk up and down stairs using suction cups installed on the soles of each feet to keep the robot attached to the stair. Jung-Yup Kim et al. [4] use force/torque sensors at ankle joint to achieve a control algorithm for a stable dynamic stair climbing. Kajita et al. [5] generated the desired COG trajectory by IPM and used a linear COG trajectory for the upward motion in stair walking.

In these methods, a humanoid robot can climb stairs successfully. But how to parameterize the procedure of climbing stairs and design stairs circumstance restriction equation need to be researched. By giving good value of parameters, the robot can climb stairs more stably and quickly.

T.-H.S. Li et al. (Eds.): FIRA 2011, CCIS 212, pp. 25–33, 2011.

This paper proposes a systematic trajectory generation method for climbing stairs. The planning of climbing stair is parameterized and stairs restrict situation is described by a series of restriction equations. First, the gait planning of climbing stairs is parameterized based on step s, walking cycle T_{sup}, height of the centre mass h_c, height of foot h_{foot}, double-support time T_{db}, rising foot time T_{up} and landing foot on the ground time T_{down}. We select suitable value of the parameters with taking account of stairs environment restriction. Then in single-support phase (SSP) hip trajectory is generated according to COG trajectory which is based on IPM in the x and y-axis. In double-support phase (DSP), hip trajectory is designed through 6th order polynomial. And ankle trajectory is planned base on position of the landing point on the ground and sine curve. Finally, according to the model parameters of SCUT-I humanoid robot, the model of the robot is built and simulated to climb stairs by matlab6.5.

2 SCUT-I Humanoid Robot Model

In order to implement the bipedal walking solution, a robot named as SCUT-I has been built, as shown in Fig.1(a).

The robot model is shown as Fig.1(b), which the arms trajectories are not taken account of. The parameters of the model are listed in Table1.

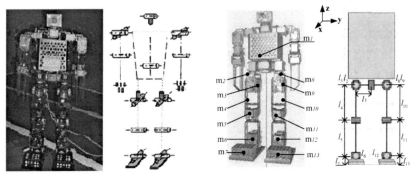

(a) SCUT-I robot   (b) SCUT-I robot model

Fig. 1. SCUT-I robot and model

Table 1. Structure parameter of SCUT-I humanoid robot

Parameter	Value	Parameter	Value	Parameter	Value	Parameter	Value
m1	1.55kg	m8	0.01kg	l_1	60mm	l_8	0mm
m2	0.01kg	m9	0.30kg	l_2	0mm	l_9	0mm
m3	0.30kg	m10	0.05kg	l_3	0mm	l_{10}	135mm
m4	0.05kg	m11	0.20kg	l_4	135mm	l_{11}	135mm
m5	0.20kg	m12	0.30kg	l_5	135mm	l_{12}	0mm
m6	0.30kg	m13	0.30kg	l_6	0mm	l_{13}	12mm
m7	0.30kg			l_7	12mm		

3   The Planning of Climbing Stairs Movement

The stair-climbing gait can be considered as a sequence of steps. Each step is composed of two phases: 1) a double-support phase (DSP) and 2) a single-support phase (SSP).

During the DSP, both feet are fixed on stairs. The robot accelerates from standing still and COG moves in the forward direction.

During the SSP, only one foot is stationary on the ground, and the other foot swings from the rear stair to the front stair. It is include three phase: 1) rising foot phase, 2) swinging foot phase and 3) landing foot phase.

The planning of climbing stairs movement is shown in Fig.2.

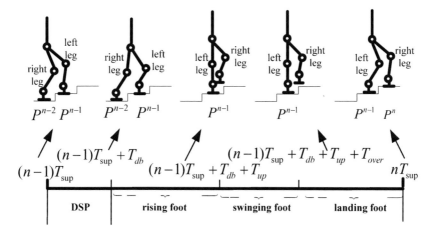

Fig. 2. The planning of climbing stairs movement

The key points of the period of cyclical climbing are defined as $(n-1)T_{\text{sup}}$, $(n-1)T_{\text{sup}}+T_{db}$, $(n-1)T_{\text{sup}}+T_{db}+T_{up}$ and $(n-1)T_{\text{sup}}+T_{db}+T_{up}+T_{over}$. Where T_{over} is swinging foot time. And $T_{down}=T_{\text{sup}}-T_{db}-T_{up}-T_{over}$.

3.1   The Parameterization of Climbing Stairs

According to the planning of climbing stairs movement, s, $T_{\text{sup}}, T_{db}, T_{up}, T_{down}, h_c$ and h_{foot} are selected to express climbing stairs movement. It is shown in equation (1).

$$P_{robot} = f(s, T_{\text{sup}}, T_{db}, T_{up}, T_{down}, h_c, h_{foot}) \qquad (1)$$

The procedure of gait planning of climbing stairs is shown in Fig.3.

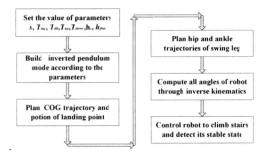

Fig. 3. The procedure of humanoid robot climbing stairs

3.2 Constraint Condition of Climbing Stairs

Because staircase is a kind of restricted walking circumstance, we need analyze its constraint situation during climbing stairs. A climbing stairs movement is shown in Fig.4. Right leg is swing leg and left leg is support leg. So in this paper we mainly analyze and design right leg's trajectory. And when left leg is swing leg, the procedure is similar.

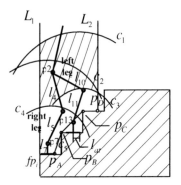

Fig. 4. A climbing stairs movement

Where p_{13} is left ankle coordinate; p_7 is right ankle coordinate; c_1 is the arc that the centre is p_{13}, radius is $l_{10} + l_{11}$; c_2 is the arc that the centre is p_7 , radius is $l_4 + l_5$; c_3 is the arc that the centre is p_{13}, radius is l_{11} ; c_4 is the arc that the centre is p_7 , radius is l_5 ; c_5 is continuous curve through P_A P_B P_C and P_D ; L_1 and L_2 are uprightness to the horizontal line. We define x_p to represent p coordinate in x-axis.

And we define $x_p^{P''}$ to represent p coordinate in x-axis at P'' position. Comparing with x, the definition of y and z are similar.

For avoiding to collide stairs, right hip coordinate p_2 should be constrained in a close curve which made of c_1, c_3, c_4, L_1, L_2. So p_2 should satisfy with equation (2).

$$gstair1: \begin{cases} x_{L_1} \leq x_{p_2} \leq x_{L_2} \\ z_{c3} \leq z_{p_2} \leq z_{c1} \cap z_{c4} \leq z_{p_2} \leq z_{c1} \end{cases} \quad (2)$$

Swing foot tiptoe trajectory should be higher than curve c_5, so $\boldsymbol{p_7}$ should satisfy with equation (3).

$$gstair2: \begin{cases} z_{p_7}(x) \geq z_{c5}(x) + l_7 \\ x_{L_1} + FL - l_{at} \leq x_{p_7} \leq x_{L_2} - l_{at} \end{cases} \quad (3)$$

Where FL is the length of foot, $z_{c5}(x)$ is shown as follows:

$$z_{c5}(x) = \begin{cases} x * \dfrac{i * SW - x_{P_A}}{SH} & i * SW - x_{P_A} \leq x \leq i * SW \\ x * \dfrac{SH}{SW} & i * SW < x \leq (i+1) * SW \end{cases}$$

Where i is number of steps, SW is the width of stair and SH is the height of stair.

3.3 The Parameters of Constraint Condition for Climbing Stairs

Firstly, s should satisfy with equation (4) for climbing stairs.

$$gstair3: SW - \frac{FL}{2} < s_x < \frac{3SW + FL}{2} \quad (4)$$

h_{foot} is at least higher than the height of stair. It should satisfy with equation (5) for climbing stairs.

$$gstair4: h_{foot} > SH \quad (5)$$

T_{up} should satisfy with equation (6) for climbing stairs.

$$gstair5: z_{tip}(t) > z_{c5\_tip}(t) \quad (6)$$

Where $z_{c5\_tip}(t)$, $z_{tip}(t)$ and $x_{tip}(t)$ are shown as follows:

$$z_{c5\_tip}(t) = \begin{cases} x_{tip}(t) * \dfrac{i * SW - x_{P_A}}{SH} & i * SW - x_{p_7} - l_{at} \leq x_{tip} \leq i * SW \\ x_{tip}(t) * \dfrac{SH}{SW} & i * SW < x_{tip} \leq (i+1) * SW \end{cases}$$

$$z_{tip}(t) = h_{foot} * \sin(\frac{t - (n-1)T_{sup} - T_{db}}{2T_{up}}\pi) + z_{p_7}^{P^{n-2}} + l_7$$

$$x_{tip}(t) = x_{p^{n-2}} + l_{at} + \frac{x_{p_7}^{P^n} - x_{p_7}^{P^{n-2}}}{T_{sg}} * (t - (n-1)T_{sup} - T_{db})$$

The setting of h_c is according to equation (2). T_{sup}, T_{db} and T_{down} should satisfy with the mechanical characteristic of robot.

4 Trajectory Generation of Climbing Stairs

If ankle and hip trajectories of swing leg are determined, all joint trajectories can be calculated by inverse kinematics; therefore, the gait of stair climbing can be planned through swing ankle joint and hip joint trajectory.

4.1 Hip Trajectory of Swing Leg

Hip trajectory is generated according to COG trajectory which is based on inverted pendulum mode.

$$p_2 = p_c + [0 \quad l_1 \quad -\Delta h_{hc}]^T \qquad (7)$$

Where p_c is COG coordinate, Δh_{hc} is the distance between COG and hip in z-axis.

$$p_c = [x_c \quad y_c \quad z_c]^T = \begin{cases} [x_{c\_dbl} \quad y_{c\_dbl} \quad z_c]^T & (n-1)T_{sup} \leq t \leq (n-1)T_{sup} + T_{db} \\ [x_{c\_sgl} \quad y_{c\_sgl} \quad z_c]^T & (n-1)T_{sup} + T_{db} \leq t \leq nT_{sup} \end{cases}$$

$[x_{c\_dbl} \quad y_{c\_dbl} \quad z_c]^T$ is COG coordinate in DSP. x_{c_dbl} and y_{c_dbl} trajectory are designed through 6th order polynomial.

$[x_{c\_sgl} \quad y_{c\_sgl} \quad z_c]^T$ is COG coordinate in SSP. x_{c_sgl} and y_{c_sgl} trajectory are designed through a 3D inverted pendulum model[6].

z_c trajectory is designed for climbing stairs and shown as follows:

$$z_c = x_c * \frac{SH}{SW} + h_c$$

Hip trajectory should satisfy with restriction condition equation (2).

4.2 Ankle Trajectory of Swing Leg

Ankle trajectory is planned base on position of the landing point on the ground. The trajectory in z-axis is planned by sine curve. The trajectories in x-axis, y-axis and z-axis are shown as follows:

$$x_{p_7}(t) = \begin{cases} x_{p_7}^{P^{n-2}} & (n-1)T_{sup} \leq t \leq (n-1)T_{sup} + T_{db} \\ \dfrac{x_{p_7}^{P^n} - x_{p_7}^{P^{n-2}}}{T_{sup} - T_{db}} * (t - (n-1)T_{sup} - T_{db}) + x_{p_7}^{P^{n-2}} & (n-1)T_{sup} + T_{db} \leq t \leq nT_{sup} \end{cases} \qquad (8)$$

$$y_{p_7}(t) = \begin{cases} y_{p_7}^{P^{n-2}} & (n-1)T_{sup} \leq t \leq (n-1)T_{sup} + T_{db} \\ \dfrac{y_{p_7}^{P^n} - y_{p_7}^{P^{n-2}}}{T_{sup} - T_{db}} * (t - (n-1)T_{sup} - T_{db}) + y_{p_7}^{P^{n-2}} & (n-1)T_{sup} + T_{db} \leq t \leq T_{sup} \end{cases} \qquad (9)$$

$$z_{p_z}(t) = \begin{cases} z_{p_z}^{p_z\;2} & (n-1)T_{\text{sup}} \le t \le (n-1)T_{\text{sup}} + T_{db} \\ h_{foot} * \sin(\dfrac{t - (n-1)T_{\text{sup}} - T_{db}}{2T_{up}}\pi) + l_7 + z_{p_z}^{p_z\;2} & (n-1)T_{\text{sup}} + T_{db} \le t \le (n-1)T_{\text{sup}} + T_{db} + T_{up} \\ h_{foot} + l_7 + z_{p_z}^{p_z\;2} & (n-1)T_{\text{sup}} + T_{db} + T_{up} \le t \le (n-1)T_{\text{sup}} + T_{db} + T_{up} + T_{over} \\ f_{h_{foot}}(t) * \cos(\dfrac{t - (n-1)T_{\text{sup}} - T_{db} - T_{up} - T_{over}}{2T_{down}}\pi) + l_7 + f_a(t) & (n-1)T_{\text{sup}} + T_{db} + T_{up} + T_{over} \le t \le nT_{\text{sup}} \end{cases}$$
(10)

Where $f_{h_{foot}} = h_{foot} - 2*SH$, $f_a = 2*SH$.

Ankle trajectory should also satisfy with restriction condition equation (3).

5 Simulation and Results

According to SCUT-I robot model, a simulator of humanoid robot climbing stairs is constructed by using Matlab6.5. A stair model is built with the length of stairs SL=0.5m, SW= 0.13m and SH=0.02m.

The parameters are given according to restriction condition and listed in Table2. COG initial coordinate is $\begin{bmatrix} 0 & 0 & h_c \end{bmatrix}^T$ and landing ground point initial coordinate is $\begin{bmatrix} 0 & s_y/2 & 0 \end{bmatrix}^T$.

Table 2. The parameters for climbing stair

T_{sup}	T_{db}	T_{up}	T_{over}	T_{down}	h_c	Δh_{hc}	h_{foot}	s_x	s_y
1s	0.2s	0.45s	0.3s	0.05s	0.3m	0.03m	0.045m	0.13m	0.12m

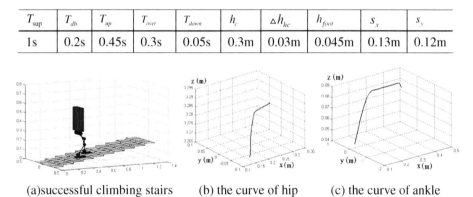

(a)successful climbing stairs   (b) the curve of hip   (c) the curve of ankle

Fig. 5. The simulation of robot's climbing stairs by using Matlab6.5

According to the planning method mentioned in this paper, the robot successfully climbed stairs. And the curves of swing ankle and hip are shown in Fig.5(b) and (c).

During robot's climbing stairs, zero moment point (ZMP) can be calculated [7]. And ΔZMP is the difference between ZMP and support area center. So in DSP ΔZMP should satisfy with equation (11).

$$\begin{cases} -(FL+s_x)/2 \le x_{\Delta ZMP} \le (FL+s_x)/2 \\ -(FW+s_y)/2 \le y_{\Delta ZMP} \le (FW+s_y)/2 \end{cases}$$
(11)

In SSP, ΔZMP should satisfy with equation (12).

$$\begin{cases} -FL/2 \le x_{\Delta ZMP} \le FL/2 \\ -FW/2 \le y_{\Delta zmp} \le FW/2 \end{cases} \tag{12}$$

Where FW is the width of robot foot. And FL = 0.14m and FW = 0.08m.

During robot's climbing stair, the curve of ΔZMP is shown in Fig.5. DSP curve is shown using solid line, SSP curve is shown using dashed line.

According to equation (11) (12), the robot can walk stably by analyzing Fig.6.ï

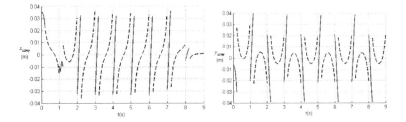

Fig. 6. The curve of ΔZMP

6   Conclusion

A desired trajectory generation method has been proposed for use throughout all the phases of robot walking stairs. The planning of climbing stair is parameterized and stairs restrict condition is described by a series of restriction equations. By setting the suitable parameters according to the restrict equations, hip and ankle trajectories are planned based on inverted pendulum model. Then a humanoid robot can climb stairs successfully and stably. This is verified in the simulation experiment.

In the future, we can optimize the parameters by using some artificial intelligence method and make robot climb stairs more stably and quickly.

References

1. Fu, C.L., Chen, K.: Gait Synthesis and Sensory Control of Stair Climbing for a Humanoid Robot. IEEE Transactions on Industrial Electronics 55(5), 2111–2120 (2008)
2. Park, C.-S., Ha, T., Kim, J.H., et al.: Trajectory Generation and Control for a Biped Robot Walking Upstairs. International Journal of Control, Automation, and Systems 8(2), 339–351 (2010)
3. Figliolini, G., Ceccarelli, M., Gioia, M.: Descending stairs with EP-WAR3 biped robot. In: Proc. IEEE/ASME Int. Conf. Adv. Intell. Mechatronics, pp. 747–752. IEEE Press, Kobe (2003)
4. Kim, J.-Y., Park, I.-W., Oh, J.-H.: Realization of Dynamic Stair Climbing for Biped Humanoid Robot Using Force/Torque Sensors. Journal of Intelligent & Robotic Systems 56, 389–423 (2009)

5. Kajita, S., Tani, K.: adaptive gait control of a biped robot based on realtime sensing of the ground profile. Autonomous Robots 4, 297–305 (1996)
6. Kajita, S., Kanehiro, F., Kaneko, K., et al.: The 3D linear inverted pendulum mode: A simple modeling for a biped walking pattern generation. In: Proceedings of IEEE/RSJ, International Conference on Intelligent Robots and Systems, pp. 239–246. IEEE Press, Maui (2001)
7. Vukobratovic, M., Borovac, B.: Zero-moment point-thirty five years of its life. International Journal of Humanoid Robotics 1, 157–173 (2004)

3D Collision-Free Trajectory Generation Using Elastic Band Technique for an Autonomous Helicopter

Chi-Tai Lee[1] and Ching-Chih Tsai[2]

[1] Aeronautical Systems Research Division,
Chung-Shan Institute of Science & Technology,
Taiwan, R.O.C.
ChiTai.Lee@gmail.com
[2] Department of Electrical Engineering,
National Chung-Hsing University,
Taichung, Twiwan, R.O.C.
cctsai@nchu.edu.tw

Abstract. A real-time path generation based on the elastic band technique is presented to find a collision-free trajectory for an autonomous small-scale helicopter flying through cluttered, dynamic three-dimensional (3D) environments. The dynamic path is followed by the adaptive trajectory tracking controller augmented with the radial basis function neural networks (RBFNN). The effectiveness and merit of the proposed method are exemplified by performing three simulation scenarios: static obstacle avoidance, dynamic obstacle avoidance and terrain following.

Keywords: elastic band, obstacle avoidance, path generation, RBFNN.

1 Introduction

Recently, researchers have paid significant attention to the study of autonomous helicopters which have outstanding advantages in a narrow environment due to their hovering, vertical take-off and landing (VTOL) maneuver abilities [1]-[10]. However, flight maneuvering in a complex environment is a major challenge for autonomous helicopters which are demanded to fly close to the ground and through cluttered urban spaces by their urban reconnaissance and survey missions. Upon reception of start and destination points from the supervisory controller, the route planner generates a feasible route for the autonomous helicopter to follow. The integrated and hierarchical approach of planning globally and reacting locally seems to provide a possible solution [10]-[12]. The so-called three-layered control system which consists of global path planning, local path re-planning, and trajectory tracking control is very suitable for an autonomous vehicle system. The middle-level of trajectory generation must work in real-time to handle local changes in the environment and to maintain a safe and smooth path followed by the low-level trajectory tracking controller. The local navigation involving with the obstacle detection and obstacle avoidance refers to the methodology of deforming the original path to overcome unexpected obstacles.

The elastic band concept was originally introduced by Quinlan and Khatib [13]-[14] for robotic path planning and recently applied to the automotive assistant systems

T.-H.S. Li et al. (Eds.): FIRA 2011, CCIS 212, pp. 34–41, 2011.

including vehicle following [15], lane-keeping, lane-changing [16], and collision avoidance [17]. It is based on the potential field approach and has been shown to provide a simple and ease computed method to reach a goal location while unknown obstacles are avoiding. Furthermore, the bubble concept offers an efficient method of implementing the elastic band as a collision-free region only if it remains inside the coverage of bubbles [10]. However, the elastic band technique has not been applied to find an admissible real-time trajectory for an autonomous helicopter maneuvering in complex 3D environments with static and dynamic obstacles.

This paper aims to present a real-time trajectory generation method for autonomous helicopters. The main contribution of this paper is that the elastic band method is successfully applied to generate a real-time, collision-free, and 3D trajectory for autonomous helicopters. The rest of the paper is organized as follows. Section 2 elaborates the complete design procedure of the real-time trajectory generation algorithm using elastic band technique with the bubble concept. In Section 3, three numerical simulations are conducted to illustrate the performance and merits of the proposed control method. Section 4 concludes the paper.

2   Real-Time Trajectory Generation Using Elastic Band Technique

The proposed trajectory generation method can be divided into four phases: initial path build-up, elastic band deformation, bubble reorganization and trajectory transformation. The detailed procedures and methods are explained in the following paragraphs.

Phase-I: Initial Path Build-up

The initial path is the shortest path between two given navigation waypoints by connecting them directly without considering any obstacles. The bubble nodes labeled with ascending numbers from 0 (start) to $N_{bubbles}$ (goal) are inserted evenly along the initial path. The bubble size gives an indication of how far the helicopter is safe from collisions. The reasonable bubble size, R_{bub}, is determined by the following equation,

$$2 \times D^{hel} = R_{min}^{bub} \leq R^{bub} \leq R_{max}^{bub} = V_{max}^{hel} \times \Delta T \tag{1}$$

where ΔT is the time interval of local path planning, D^{hel} is the diameter of the main rotor and V_{max}^{hel} is the maximum speed of the unmanned helicopter. The total number of bubble nodes is calculated by $N_{bubbles}$ = ceil(D_{min}/R_{max}^{bub}), where D_{min} is the shortest distance from the start position to the goal position.

Phase-II: Elastic Band Deformation

The initial path is continuously modified in real-time according to the latest information of the environmental static and dynamic obstacles. Two virtual forces are introduced to describe the interaction between bubbles or with external obstacles. Each node (b_i) is attracted by two internal forces from its preceding node (b_{i-1}) and following node (b_{i+1}) respectively. The internal force for a bubble b_i is computed using the following equation.

$$F_{int}^{bi} = k_{int} \left[\frac{\vec{b}_{i+1} - \vec{b}_i}{\|\vec{b}_{i+1} - \vec{b}_i\|} \left(\|\vec{b}_{i+1} - \vec{b}_i\| - R_{min}^{bub} \right) + \frac{\vec{b}_{i-1} - \vec{b}_i}{\|\vec{b}_{i-1} - \vec{b}_i\|} \left(\|\vec{b}_{i-1} - \vec{b}_i\| - R_{min}^{bub} \right) \right] \quad (2)$$

where $\|\cdot\|$ denotes the Euclidean norm and k_{int} means the contraction gain. The external forces are repulsive forces exerted by the obstacles of the environment to deform the bubble band adequately and then keep the path collision-free. Each node (b_i) is repelled by all the nearby obstacles only if they are close enough. The single repulsive force from one of obstacle is calculated as follows:

$$\vec{f}_{ext}^{\,j}(b_i) = k_{ext} e^{-(D_{aff})} \left(\frac{\vec{b}_i - \vec{O}_j}{\|\vec{b}_i - \vec{O}_j\|} \right) \quad (3)$$

where k_{ext} means the repulsive gain. The fading function depends on the affected distance D_{aff} which is calculated via

$$D_{aff} = \|\vec{b}_i - \vec{O}_i\| - D_{safe}, \; if \; \|\vec{b}_i - \vec{O}_i\| > D_{safe}; \; zero \; otherwise. \quad (4)$$

where the safe distance is defined by $D_{safe} = R_{min}^{bub} + V_{bub}^{obs} \times T_s$ and V_{bub}^{obs} is the relative velocity from the obstacle to this bubble. Finally, the resultant external force acting on this bubble node is calculated by

$$\vec{F}_{ext}^{bi} = \sum_{j=1}^{N} \vec{f}_{ext}^{\,j}(b_i) \quad (5)$$

where N is the total number of nearby obstacles. The net applied force for the node b_i is summarized by

$$\vec{F}_{net}^{bi} = \alpha \cdot \vec{F}_{int}^{bi} + \beta \cdot \vec{F}_{ext}^{bi} \quad (6)$$

where α and β are respectively the weighting factors for the internal forces and external forces. The new elastic band configuration for each bubble node is

$$\vec{P}_{new}^{bi} = \vec{P}_{old}^{bi} + \gamma \cdot \vec{F}_{net}^{bi} \quad (7)$$

where γ is the step size for updating the bubble band deformation.

Phase-III: Bubble Reorganization

The two following properties of the elastic band must be checked properly not only to maintain it as a continuous and feasible path, but also to improve its efficiency. First, it is desirable to remove redundant bubbles from elastic band when a small bubble's coverage is totally within its neighboring big bubble's coverage. Thus, the k-th bubble is removed if the criterion for bubble redundancy defined by (8) is true,

$$|R_k - R_{k-1}| \geq \|\vec{b}_k - \vec{b}_{k-1}\| \; or \; R_{k+1} + R_{k-1} > \|\vec{b}_k - \vec{b}_{k-1}\| + \|\vec{b}_{k+1} - \vec{b}_k\| \quad (8)$$

where $|\cdot|$ means the absolute value and $\|\cdot\|$ is the 2-norm to measure the distance. Second, there is bubble insufficiency when two neighboring bubbles do not overlap each other due to the bubble moving too far during the path deformation. If the criterion for bubble insufficiency defined in (9) is true, then an extra bubble needs to be inserted at the middle location between b_{k-1} and b_k bubbles.

$$R_k + R_{k-1} - d_{ol} < \left\| \vec{b}_k - \vec{b}_{k-1} \right\| \tag{9}$$

where d_{ol} is the desired overlap distance between two neighboring bubbles.

Phase-IV: Trajectory Transformation

Finally, this feasible path generated from the elastic-band planner is smoothed by using the cubic B-spline technique. The nodes of the elastic band will be interpolated by a cubic spline curve using the Matlab Spline Toolbox, where $\Gamma_{EB} = $ cscvn (nodes) is a parametric cubic spline curve through given points. The vehicle's velocity is controlled by passing through those nodes at specified time interval as follows,

$$\vec{V}_i = \frac{\vec{b}_{i+1} - \vec{b}_i}{T_c} \tag{10}$$

The closer the bubble nodes allocation, the slower the helicopter flight speeds.

Real-Time Algorithm for an Elastic Band Trajectory Generation

On basis of the aforementioned descriptions, a real-time algorithm for elastic band trajectory generation is proposed in the following steps. This 16-step procedure is executed iteratively from the current position to the goal waypoint.

1: Construct the initial path Γ_{ini} consisting of $N_{bubbles}$ bubbles allocated evenly between the start waypoint and the goal waypoint. Refer to Phase-I.
2: Perform the elastic band deformation process from the current bubble node through all the uncompleted bubble nodes.
3:    Compute the internal forces acting on the i-th bubble node due to its neighbor nodes of index $i-1$ and $i+1$ respectively. Refer to (2).
4:    Calculate the external forces acting on the i-th bubble node due to all the static and dynamic obstacles. Refer to (3)-(5).
5:    Execute the path deformation for the i-th bubble node applying the sum of the virtual forces given in Steps 3 and 4. Refer to (6)-(7).
6:    Repeat Steps 3-5 until the deformation amount is less than the tolerance.
7:    Decide bubble's radius on the smallest clearance distance to all the obstacles.
8:    Shift the bubble index to the next and repeat Steps 3 to 7 until the end of the elastic band, $i = N_{bubbles}$.
9: Do the bubble reorganization from the current bubble node through all the uncompleted bubble nodes.
10:    Check if the i-th bubble is redundant and deletes it if (8) is true.

11:   Check if the i-th bubble has a broken connection with its previous one and in-
       serts an extra bubble at the middle of two bubbles if (9) is true.
12:   Do the path deformation if an extra bubble is generated. Refer to Steps 3-6.
13:   Shift the bubble index to the next and repeat Steps 10 to 12 until the end of the
       elastic band, $i = N_{bubbles} - 1$.
14: Generate the collision-free trajectory from the current position to the goal.
15:   Smooth the deformed elastic band path Γ_{EB} by using a cubic spline function.
16:   Generate a real-time trajectory by correlating the path Γ_{EB} with time by (10).
      Repeat Steps 2-16 until the goal is achieved ($i = N_{bubbles}$).

3   Simulations and Discussion

In this section, two numerical simulations are conducted to investigate the effectiveness
and performance of the proposed method which was implemented as the middle-level
local path planner to find a feasible collision-free trajectory. The flight trajectory is
followed by the low-level motion controller which adopts an intelligent trajectory
tracking method augmented with a RBFNN approximation to the coupling forces
between the helicopter actuators [6]. The missions of an autonomous helicopter are to
fly through the 3D complex environments and arrive at the specified goal position
given by an off-line global path planner. Simulation cases include the dynamic obstacle
avoidance and the terrain following.

3.1 Dynamic Obstacle Avoidance

For applications of ground robots, the elastic band path can not cope with those cases
where external obstacles move across it. However, this case intends to execute the
extreme case for an autonomous helicopter flying through a 3D environment. The
scenario is described first of all. Both starting and goal waypoints are given in 3-D
coordinates as [0 0 0] and [100 100 30] (unit: meters), respectively. There are four static
obstacles blocking the direct path from the start-up to the goal in the scattered environment. They are chosen arbitrarily as follows: [15 10 40 5], [20 25 30 5], [50 50 5 5],
[85 80 45 5], where each bracket contains 3D coordinates and radius in meters. In
addition, a moving obstacle of 10-meter radius starts at [40 15 10] meters and is going
across the original trajectory with a velocity of [-2 2.6 0] meters/second. In the virtual
3D environment, all static obstacles are represented by cylinder objects with different
heights and radii and the dynamic obstacle is represented by a ball object. Furthermore,
the minimum bubble radius (R_{min}^{bub}) and maximum bubble radius (R_{max}^{bub}) are limited by 1.6
and 10 meters. The time interval for local trajectory planning is 1 second, and the
update rate for trajectory tracking controller is 50 Hertz.
      Fig. 1(a) depicts that the arising elastic-band path deformed downward to avoid
collision with the moving obstacle. Affected by the right-hand coming obstacle, the
elastic band was repelled to deform toward the left-hand side. Up to the extreme length
sustained by the internal forces of the elastic band, its deformation type changed from
stretching to rotation. Taking advantage of the 3D environment, the proposed method

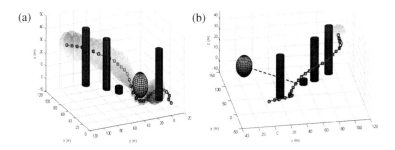

Fig. 1. Simulation results of collision-free flight trajectories among 3D virtual environment scattered with static and dynamic obstacles

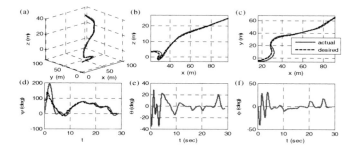

Fig. 2. Time responses of the control outputs and attitude angles of an autonomous helicopter while the simulation in Fig.1.

finds out which movement direction can reduce the threat of external obstacles. As shown in Fig. 1(b), the elastic-band trajectory passing under the moving obstacle is totally collision free. After passing over the external dynamic object, the elastic band arose quickly back to the desired height and led to the goal waypoint. Figs. 2(a)-2(f) depict that the desired trajectories of control outputs are tracked well except for tight turns and abrupt direction changes. Therefore, the elastic band method in 3D environment can cope with the cases of external obstacles moving across the desired path.

3.2 Terrain Following

The proposed method is suitable for not only obstacle avoidance but also terrain following which is another important requirement for an autonomous helicopter. The scenario for the third simulation case is described as follows. The 3D terrain model is created with ascent and descent hillsides whose climax is around 24 meters. The start and goal waypoints are given in X-Y coordinates as [-40 40] and [40 -40] meters, respectively. The height is desired to be 15 meters above the ground. The minimum bubble radius (R_{min}^{bub}) and maximum bubble radius (R_{max}^{bub}) are limited by 1.6 and 10 meters respectively. The time interval for local trajectory generation is 1 second. The update rate for trajectory tracking controller is 50 Hertz. The simulation results are depicted in Fig. 3 and Fig. 4.

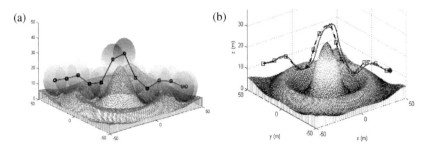

Fig. 3. Simulation results of terrain following flight trajectories among 3D virtual environment

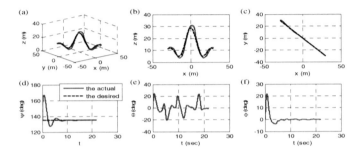

Fig. 4. Time responses of the control outputs and attitude angles of an autonomous helicopter while the simulation in Fig.3

As shown in Fig. 3(a), the initial path was also constructed as a straight line to directly cross the start and the goal waypoints. After interaction with the terrain data, the elastic band deformed according to the ascending and descending terrain height. Furthermore, the elastic band executed the bubble reorganization to improve bubble redundancy and bubble insufficiency. Fig. 3(a) depicts that this deformed path is safe to be collision-free from the terrain threat and feasible to be a continuous path for a motion controller. The path deformation and reorganization are iterated sequentially from current bubble node to the goal waypoint. The elastic band is gradually shaped as the blue path shown in Fig. 3(b). Based on the elastic band, the real-time trajectory is generated piece by piece between two neighboring bubble nodes. Finally, the red line in Fig. 3(b) denotes the actual trajectory executed by the proposed intelligent adaptive trajectory tracking controller. Its tracking performance is further examined in Fig.4. As shown in Fig. 4(b), the trajectory deviation is mainly caused by the height variation which is a little but still clear from the terrain. Fig. 4(d) depicts that the heading control is disturbed in the beginning but has a good convergence property. The attitude variations of around 20 degrees in Figs. 4(e)–4(f) means that the proposed controller has a good maneuverability for terrain following.

4 Conclusion

This paper has proposed a real-time path generation method which is based on the elastic band technique to find a collision-free trajectory for an autonomous helicopter. The detailed procedure is synthesized as a 16-step routine and executed iteratively till the goal position is reached. The hybrid hierarchical control system is constructed by combing the proposed local path planner and the intelligent trajectory tracking controller augmented with a RBFNN approximation [6]. The effectiveness and merit of the proposed control system have been exemplified by performing three simulations: static obstacle avoidance, dynamic obstacle avoidance and terrain following. From the three simulation results using the high-fidelity and well-validated dynamic model, the proposed method has been shown not only safe and feasible for a wide range of helicopter flight conditions, but also satisfactory in obtaining transient performance of flight maneuvers.

References

1. Gavrilets, V., Mettler, B., Feron, E.: Nonlinear model for a small-size acrobatic helicopter. In: Proc. of AIAA Guidance, Navigation and Control Conference, AIAA 2001, vol. 4333 (2001)
2. Mettler, B.: Identification modeling and characteristics of miniature rotorcraft. Springer, Heidelberg (2003)
3. Mahony, R., Hamel, T.: Robust trajectory tracking for a scale model autonomous helicopter. Int. J. of Robust and Nonlinear Control 14, 1035–1059 (2004)
4. Johnson, E.N., Kannan, S.K.: Adaptive Trajectory control for autonomous helicopters. AIAA J. of Guidance, Control, and Dynamics 28, 524–538 (2005)
5. Marconi, L., Naldi, R.: Aggressive control of helicopters in presence of parametric and dynamical uncertainties. Mechatronics 18(7), 381–389 (2008)
6. Lee, C.T., Tsai, C.C.: Improved nonlinear trajectory tracking using RBFNN for a robotic helicopter. Int. J. of Robust and Nonlinear Control, 1–18 (2009), doi:10.1002/rnc.1483
7. Lee, C.T., Tsai, C.C.: Real-Time Path Planning for Autonomous Control of a Small-Scale Helicopter. In: Proc. of 2009 National Symposium on System Science and Engineering, Tamsui, Taiwan, June 26 (2009)
8. Lee, C.T., Tsai, C.C.: Adaptive Backstepping Integral Control of a Small-Scale Helicopter for Airdrop Missions. Asian Journal of Control 12(4), 531–541 (2010)
9. Lee, C.T., Tsai, C.C.: Nonlinear Adaptive Aggressive Control Using Recurrent Neural Networks for a Small Scale Helicopter. Mechatronics 20(4), 474–484 (2010)
10. Lee, C.T.: Trajectory Planning and Adaptive Trajectory Tracking Control for a Small Scale Autonomous Helicopter. Ph.D. (2010)
11. Vachtsevanos, G., Tang, L., Drozeski, G., Gutierrez, L.: From mission planning to flight control of unmanned aerial vehicles: Strategies and implementation tools. Annual Reviews in Control 29, 101–115 (2005)
12. Yang, H., Zhao, Y.: Trajectory Planning for Autonomous Aerospace Vehicles amid Known Obstacles and Conflicts. J. of Guidance, Control, and Dynamics 27, 997–1008 (2004)
13. Quinlan, S., Khatib, O.: Elastic bands: Connecting path planning and control. In: Proc. IEEE Conf. Robot. Autom., pp. 802–807 (1993)
14. Khatib, O.: Real-time obstacle avoidance for manipulators and mobile robots. International Journal of Robotics Research 5(1), 90–98 (1995)
15. Hilgert, J., Hirsch, K., Bertram, T., Hiller, M.: Emergency Path Planning for Autonomous Vehicles Using Elastic Band Theory. In: IEEE/ASME International Conference on Advanced Intelligent Mechatronics, Kobe, Japan, July 20-24 (2003)
16. Gehring, S.K., Stein, F.J.: Elastic Bands to Enhance Vehicle Following. In: IEEE Intelligent Transportation Systems Conference Proceedings, Oakland (CA), USA (2001)
17. Hesse, T., Sattel, T.: An Approach to Integrate Vehicle Dynamics in Motion Planning for Advanced Driver Assistant Systems. In: Proc. of 2007 IEEE Intelligent Vehicle Symposium, Istanbul, Turkey, June 13-15 (2007)

A Versatile Kit for Teaching Intelligent Mobile Robots

Juing-Huei Su, Chyi-Shyong Lee, Hsin-Hsiung Huang,
Shih-Wei Chao, Sheng-Hong Lin, and Yu-Cheng Wu

Dept. of Electronic Engineering, Lunghwa University of Science and Technology,
No. 300, Sec. 1, Wan-Shou Rd., Touyuan county, Taiwan
suhu@mail.lhu.edu.tw

Abstract. The development of a versatile kit to raise student interest in learning the implementation skills of intelligent mobile robots is presented in this paper. The kit is capable of solving micromouse mazes, line mazes, and following line tracks with different curvatures at different speed settings. It is first devised to be used in various project-oriented hands-on laboratory courses for students in the department of electronic engineering of Lunghwa University of Science and Technology, and introductory workshops for vocational high school students and teachers with electronic and information engineering backgrounds. To enhance the learning outcomes, contests can also be organized for students to see how well the techniques learned in the laboratory are applied in their mobile robots.

Keywords: robot education, intelligent mobile robots.

1 Introduction

Robots are attracting more and more people's attention recently [1], especially when iRobot announced the Roomba vacuum cleaning robot [2]. This makes the fundamental understanding of robots a necessity for many electronic engineers [3]. Unfortunately, learning the design philosophy of robots is interesting but difficult, because it includes many areas of knowledge, e.g., mechanics and electronics, automatic control theory, and software programming of microcontrollers, etc. It is found that students will be willing to do tedious research work to solve practical problems when these problems are related to an interesting, and competitive contest [4]. Winning one or two awards at a competition not only gives students a sense of accomplishment but also gives pride and visibility to their schools. This is an important factor for technology oriented university students in Taiwan, because of their low learning achievements in traditional theory oriented lecture courses.

One of the problems with commercially available mobile robot kits [5-6] for autonomous mobile robot courses is that they are usually expensive. Laboratory kits also require frequent maintenance services as a result of component or module failures. They do become obsolete quite rapidly as new products or technologies are developed. Therefore, a low-cost and versatile intelligent mobile robot kit is first devised in the Lunghwa University of Science and Technology to help not only raise student interest, but also motivate them to learn actively the implementation skills and related theories about autonomous mobile robots.

T.-H.S. Li et al. (Eds.): FIRA 2011, CCIS 212, pp. 42–49, 2011.
© Springer-Verlag Berlin Heidelberg 2011

Project-based laboratory exercises based on the kit are also developed to lead students step by step in making the mobile robot search successfully the goal square of the maze, memorize the curvatures of line tracks, and plan different speed settings for fast runs. The contents of the hands-on laboratory includes 1) introduction of the integrated development environment (IDE) for the kit, 2) interrupt driven real-time control firmware structure, 3) implementation of a simple maze-solving algorithm, 4) calibration of optical sensors, 5) maze wall detection via infrared light emitting diodes (LEDs) and optical sensors, 6) dc servo motor control with home-made encoders, 7) position estimation of line tracks via interpolation techniques, and 8) position and orientation contorl of the mobile robot. The hands-on laboratory exercises are described in details as follows.

2   The Versatile Robot Kit

The kit shown in Fig. 1 is steered with dc motors and differential drive. Two simple encoder circuits with resolution of 12 pulses per revolution are also designed by using optosensors.   The firmware can be downloaded to the flash memory of the microcontroller dsPIC33FMC804 through an in system programming (ISP) port. Students can collect the data stored in the microcontroller, or send commands to the firmware via an RS232 serial bus.   When the robot is used as a micromouse, it controls 6 infrared light emitting diodes (LEDs) in 5 directions, and detects the intensity of the reflected light to determine the maze wall information and to correct the motion commands for two dc motors. The kit can also keep track of its position in the maze by using the pulses which come from the encoder cirucuits.   The firmware in the microcontroller can interact with the user with buttons and matrix LED display. By using the functions the mircromouse kit provides, students should devise their own maze-solving algorithm to help the micromouse find out the goal and decide an optimal route from the start to the goal according to its motion capability.

The followings describe how the versatile robot kit is used in various parts of the hands on laboratory in leading the students to learn intelligent mobile robots.

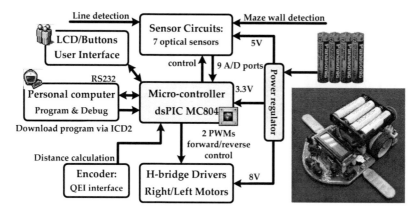

Fig. 1. The low cost intelligent mobile robot kit for hands on laboratory courses

2.1 An Interrupt Driven Real-Time Control Firmware

The mobile robot has to calculate its position according to its encoders and scan the environment with the infrared LEDs and optical sensors constantly, such that the errors in position calculations can be corrected. This relies heavily on an interrupt driven real-time control firmware structure. Students would learn in this part how to control the time intervals of system functions for peripherals, motion control, and the maze solving algorithm. These time-intervals should be short enough to make the mobile robot react fast enough to environment changes, which is important when the mobile robot is running fast. Key factors to influence the code execution efficiency, such as fixed-point mathematics, table-lookup skills for trigonometric functions, and programming skills would be introduced to students in this part. The corresponding laboratory exercises will also be conducted.

2.2 A Simple Maze-Solving Algorithm

The simplest method for a mobile robot to solve given mazes is some variation on the flood-fill or Bellman algorithm [7]. The idea is to start at the goal square and fill the maze with values which represent the distance from each square in the maze to the goal square. When the flood reaches the start square, the algorithm can then be stopped. The mobile robot could follow the values downhill to the goal square. Although the mobile robot knows nothing but the start square about the maze configuration at first, it can still follow a route suggested by the flood algorithm if unvisited squares are assumed to contain no maze walls. When entering an unvisited square, it records the wall information and remembers the visited status of that square. By using the procedure described above, the goal square would be found at last.

When the goal square and the shortest paths from the start square to the goal square are found, the shortest paths can be furthermore optimized such that the mobile robot can run in diagonal instead of consecutive 90 degree turns to reduce even more the run time from the start square to the goal square.

25	24	23	22	21	20	11	10	9	6	5	4	3	2	1	G
26	25	24	23	20	19	12	9	8	7	6	5	4	3	2	1
27	26	25	22	19	18	13	14		8	9	6	5	4	3	2
28	27	22	21	18	17	16	15	10	9	8	7	8	5	6	3
27	24	23	20	19	18	17	14	11	12	9	18	9	6	7	4
26	25	20	19	18	17	16	13	12	11	10	17	16	9	8	5
27	22	21	18	17	16	15	14	15	16	19	18	15	10	11	6
S	23	20	19	18	19	18	17	16	17	18	19	14	13	12	7

Fig. 2. Optimization of the shortest path (in blue) with diagonal running path (in green)

2.3 Calibration of Optical Sensor Outputs

To detect the maze wall or the line in the racing track, most people use reflective optical sensors. Threshold values of the reflected light intensity can be used to determine whether or not the maze wall exists. If one wants to get more accurate estimation of the distance between robot and the maze wall, or the position of the line with respect to the center point of the robot, analog output voltages from reflective optical sensors should be used. Before students can use the analog sensor outputs to estimate line positions or maze wall distance, each output should be calibrated to give almost the same level of voltage under the same working condition. This is due to the variations of optical sensor characteristics, even though their part numbers are the same. It can be seen in figure 3a that the output values of different reflective optical sensors vary a lot even if they are under the same working condition. Figure 3b shows the results after the software calibration procedure [8] is applied.

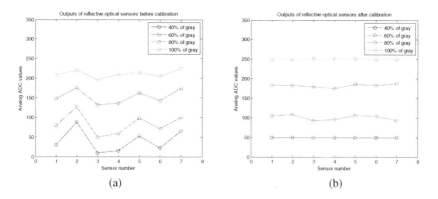

(a)   (b)

Fig. 3. The optical sensor outputs for different gray scale (a) before and (b) after calibration

2.4 Line Detection via Interpolation Techniques

After the calibration of optical sensor outputs is finished, it is ready to calculate the line position based on those adjusted analog output values. Students can use various interpolation techniques or the way of find the "center of mass" [9] to do it. It is important at this part for students to figure out a way to verify how accurate their estimation algorithm is.

Suppose that the coordinate of the 4 reflective optical sensors are x_0, x_1, x_2, x_3, x_4, x_5, x_6, respectively, and the corresponding analog output values are y_0, y_1, y_2, y_3, y_4, y_5, y_6, which is shown in figure 4. The estimated line position can then be calculated by the following weighted average formula:

$$x = \frac{\sum_{i=0}^{7} x_i y_i}{\sum_{i=0}^{7} y_i} = \frac{3(y_6 - y_0) + 2(y_5 - y_1) + (y_4 - y_2)}{\sum_{i=0}^{7} y_i}, \qquad (1)$$

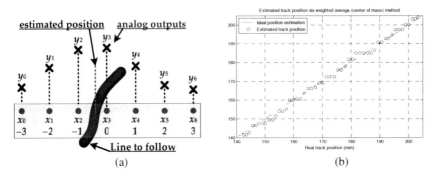

(a)   (b)

Fig. 4. The experimental verification of the center of mass line detection algorithm

2.5 Motion Control of dc Servo Motors with Home-Made Encoders

To reduce the total cost of the mobile robot kit for education without sacrificing too much of its capabilities, toy dc motors are used with home-made 12 pulse encoders (shown in Fig. 5) are used to control the speed and position of the robot.

Although position feedback control loop is used in implementing the motion control algorithms, the performance suffers from the low resolution of the encoder. It is shown in the upper half of Fig. 6 that the position is poorly controlled because of the derivative of error between command and output signals can not be accurately calculated (shown in the lower left part of Fig. 6 as a green curve). Therefore, an efficient velocity estimation method, whose block diagram is shown in Fig. 7, is applied to obtain a better signal when compared with traditional difference method. This is also shown in the lower left part of Fig. 6 as a red curve. It can be easily seen that the method do improve the estimation of velocity signal. In addition to estimating the velocity signal from low resolution encoder outputs, the method presented in Fig. 7 can also estimate the acceleration signal. It may be used in feedforward control method in the future for even better performance.

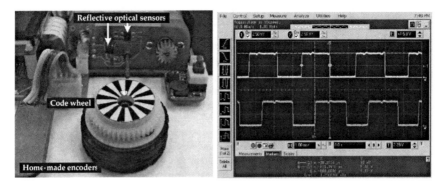

Fig. 5. The home-made encoder and its corresponding 2 phase signals

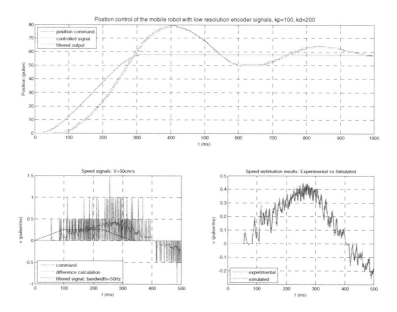

Fig. 6. The experimental results of controlling the position of the mobile robot via low resolution encoder signals

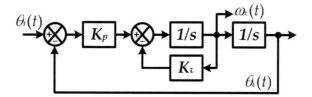

Fig. 7. An efficient method for estimating velocity from low resolution encoder signals

The experimental results in Fig. 8 clearly show the improved performance in velocity control when the velocity estimation method is applied in the feedback control loop.

3   Micromouse, Robotrace and Line Maze Contests

The low cost robot education kit presented in this paper is capable of participating in contests about micromouse, line following, and solving line mazes, which are shown in Fig. 9a-c, respectively. The contests not only keeps learning interest of students, but also drive them to explore in more details the implementation skills about mobile robots. Therefore, it is really cost-effective for students to learn sensor characteristics calibration, motion control, maze solving algorithms, line detection and following algorithms, and their firmware implementations.

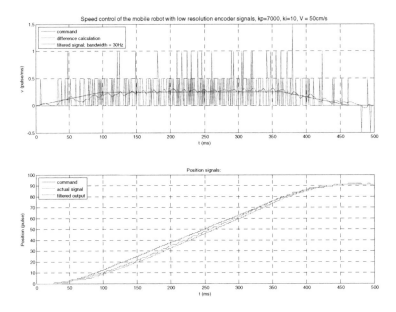

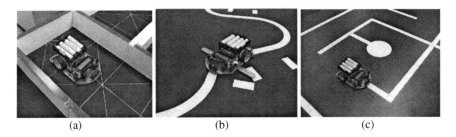

Fig. 8. The experimental results of controlling the position of the mobile robot via an estimated velocity signal

Fig. 9. The (a) micromouse, (b) line-following, and (c) line-maze contests for the mobile robot presented in this paper

4   Conclusions

A low cost but versatile mobile kit first devised in Lunghwa University of Science and Technology and used as a learning platform for autonomous robots is presented in this paper. Several project-oriented laboratory exercises are also devised for the kit to help students learn the basics about mobile robots. They include 1) introduction of the integrated development environment (IDE), 2) interrupt driven real-time control firmware structure, 3) implementation of a simple maze-solving algorithm, 4) maze wall detection via infrared light emitting diodes (LEDs) and optical sensors, 5) motion control of dc motors with low resolution encoders, 6) line detection via interpolation techniques, and 7) calibration of sensor characteristics. The kit is designed with help from the local branch of Microchip Inc., therefore, students can get the necessary C-compilers and microcontrollers free of charge. It is hoped that every student can have his/her own mobile robot for learning implementation skills easily even if he/she is economically disadvantaged, and learn quickly the techniques in making an autonomous mobile robot. The feedback from students shows that the kit and related national contests do effectively raise students' interest, and therefore motivate them to learn actively those related theory and implementation skills. The overall cost of the kit is less than USD$70, which is well below the cost of similar commercially available kits [9-10].

Acknowledgments. The authors would like to thank the National Science Council for its financial support under grant NSC 99-2516-S-262-002, and Microchip Inc. for their support of free C-compilers and microcontrollers.

References

1. Gates, B.: A robot in every home. Scientific American Magazine (January 2007)
2. iRobot cooperation, Vacuum cleaning robot - iRobot Roomba (2009),
   http://www.irobot.com (accessed June 21)
3. Qu, Z., Wu, X.: A new curriculum on planning and cooperative control of autonomous mobile robots. International Journal of Engineering Education 22(4), 804–814 (2006)
4. Chen, N., Chung, H., Kwon, Y.K.: Integration of micromouse project with undergraduate curriculum: a large-scale student participation approach. IEEE Transactions on Education 38(2), 136–144 (1995)
5. Micromouse AIRAT2, http://www.robotstorehk.com/micromouse/RS-AIRAT2.html
6. LEGO mindstorm, http://mindstorms.lego.com/en-us/default.aspx
7. Society of Robots, Maze Solver,
   http://www.societyofrobots.com/member_tutorials/node/94
8. Lee, C.-S., Su, J.-H., Lin, K.-E., Chang, J.-H., Chiu, M.-H., Lin, G.-H.: A hands-on laboratory for autonomous mobile robot design courses. In: The 17th Int. Federation of Automatic Control World Congress, Seoul, Korea, July 6-11, pp. 9473–9478 (2008)
9. Active Robots, AIRAT2 MicroMouse Maze Solving Robots, http://www.active-robots.com/products/robots/maze-details-1.shtml
10. Pololu, 3pi Robot, http://www.pololu.com/catalog/product/975

A Novel Approach of Robust Active Compliance for Robot Fingers

Jamaludin Jalani[1,2], Said Ghani Khan[2,3],
Guido Herrmann[1,2], and Chris Melhuish[1,2,3]

[1] Department of Mechanical Engineering, University of Bristol, Queen's Building
University Walk Bristol BS8 1TR UK
[2] Bristol Robotics Laboratory (BRL) DuPont Building Bristol Business Park Bristol
BS16 1QD UK
[3] University of the West of England Coldharbour Lane Frenchay Bristol BS16 1QY
UK

Abstract. In order to guarantee that grasping with robot fingers are safe when interacting with a human or a touched object, the robot fingers have to be compliant. In this study, a novel active and robust compliant control technique is proposed by employing an Integral Sliding Mode Control (ISMC). The ISMC allows us to use a model reference approach for which a virtual mass-spring damper can be introduced to enable compliant control. The performance of the ISMC is validated for the constrained underactuated BERUL (Bristol Elumotion Robot fingers) fingers. The results show that the approach is feasible for compliance interaction with objects of different softness. Moreover, the compliance results show that the ISMC is robust towards nonlinearities and uncertainties in the robot fingers in particular friction and stiction.

Keywords: Integral Sliding Mode Controller, Active Compliance Control, Underactuated Robot Fingers, Robust Control.

1 Introduction

The capability of human fingers to grasp any object without damage is a fundamental part in achieving safe human robot interaction [1], [2]. It is expected that robots are going to be used in a care taking and home environment [3], which requires them to be particularly safe. Compliance in this case is very important and has to encompass compliance which is usually observed in an industrial context, i.e. spot welding or cargo carrying [4]. Compliant design approaches include passive compliance [5], [6], [7], [8], [9], active compliance control [10], [11], [12], [13] and hybrid compliance control [14]. Moreover, the kinematics of the robot fingers can be essential to achieve compliance and suitable grasping [15]. In case, robot fingers are underactuated, compliance is not easily achieved and the kinematics of the fingers are important [16].

There has been recently a significant amount of work in the area of active compliance. In particular, Wang *et al.* [4] argue that active compliance is easily realized. The work in [10], [11], [12] and [13] proves that active compliance

T.-H.S. Li et al. (Eds.): FIRA 2011, CCIS 212, pp. 50–57, 2011.

control via a simple PD control approach is possible, although in many cases a robust controller method is preferable. For this, an appropriate controller scheme, which allows for model reference to achieve compliance and to provide robust control, is desirable for robot fingers [17]. It is expected the selected controller will eliminate or at least mitigate the nonlinearities and uncertainties.

The existence of friction and stiction can degrade control performance [18], [19]. For this, it must be noted that the exact friction model is difficult to be obtained in practice. In [17], we argue that robust sliding mode control is particularly suitable for the control of robotics fingers subject to friction. For this purpose, an integral sliding mode controller (ISMC) introduced by Shi [20] is proposed and investigated in our case. In contrast to [17], we suggest and investigate practically the introduction of an externally measured force signal into the reference model of the integral sliding mode controller to achieve active compliance.

Despite ISMC appears to be very useful due to its model reference behavior, the use of the ISMC in real time applications is still on the rise and only a small number of ISMC applications are reported. Recent studies closely related to the application of the BERUL fingers, can be seen in the control of a two-link rigid manipulator by comparing SMC and ISMC in simulation [20]. Other applied work of ISMC has been recently provided by [21] for a power system. The work in [22] employs ISMC in a unicycle type mobile robot while the authors in [23], implement ISMC in an electromechanical system and the work in [24] combines ISMC and H_∞ control in a simulation. However, recently, the authors conducted first tests using the ISMC for the BERUL fingers, showing excellent performance when rejecting friction and stiction [17].

2   Controller Design

For controller design, we are using robot fingers where the joints are moved together using a push rod [17], [25]. This allows to model each finger as.

$$m\ddot{q} + f = u \tag{1}$$

where m is the generalized mass/inertia, f is a lumped expression for the major nonlinearities i.e. gravity, friction and centrifugal/coriolis force. In addition, the above simplified model is possible due the robust method used here.

2.1   Integral Sliding Mode Controller (ISMC)

Motivated by the recent development of the ISMC on a two-link rigid robotic manipulator [20] through simulation, the same controller is applied on the BERUL fingers. Using the general dynamic equation of (1), a suitable integral sliding mode controller [20] is designed as follows: The joint torque vector τ can be split into two additive terms:

$$\tau = \tau_0 + \tau_1 \tag{2}$$

$$\tau_0 = m_0(q)(\ddot{q}_d - K_D\dot{q}_e - K_P q_e + f_0(q,\dot{q})), \tag{3}$$

where $m_0(q)$, $f_0(q, \dot{q})$ are the nominal values of $m(q)$ and $f(q, \dot{q})$ respectively; which are defined as $m_0(q) = m(q) - \Delta m$ and $f_0(q) = f(q) - \Delta f$; $K_P \in \Re^{n x n}$ and $K_D \in \Re^{n x n}$ are positive scalars determining the closed loop performance; and the tracking error is defined as $q_e(t) = q(t) - q_d(t)$ with $[q_d(t) \; \dot{q}_d(t) \; \ddot{q}_d(t)]$ being the reference trajectory and its time derivatives. Note that τ_0 represents a feedback linearization component with PD control and τ_1 denotes a discontinuous torque control.

The discontinuous torque control and sliding manifold are respectively defined as

$$\tau_1 = -\Gamma_0 sign(s) \tag{4}$$

$$s = \dot{q}_e + K_s q_e + K_i \int_0^t q_e(\xi) d\xi - \int_0^t G_f H - \dot{q}_e(0) - K_s q_e(0) \tag{5}$$

where G_f is positive scalar and H is an external force measurement, obtained via specially introduced sensors. K_s is a damping coefficient and K_i is a stiffness coefficient. The combination of $(\tau_0 + \tau_1)$ create a robust ISMC-controller, which, however, may cause chattering in the control signal. In order to avoid the effect of the chattering, the following equation is used,

$$\frac{s}{|s| + \delta} \tag{6}$$

instead of sign(s) where the scalar $\delta > 0$ allows to suppress chattering.

2.2 Compliance

For compliance, we reconsider the sliding variable s and its derivative:

$$\dot{s} = \ddot{q}_e + K_s \dot{q}_e + K_i q_e - G_f H \tag{7}$$

When sliding mode is achieved, then $s = 0$ and in particular $\dot{s} = 0$. For $\dot{s} = 0$, the error dynamics are defined by the damping constant K_s, the spring constant K_i and the external force measurement signal H introduced via the input distribution gain G_f, namely

$$\ddot{q}_e + K_s \dot{q}_e + K_i q_e = G_f H \tag{8}$$

This defines a reference model allowing for active compliance control and contrasts to the recent of ISMC, where the sliding mode dynamics are generally defining a nominal closed loop behavior without external signals. This is an important tool as the controller guarantees a well defined level of compliance despite the high degree of uncertainty and friction in the robot hands. A virtual reference model for this is

$$\ddot{q}_r = -K_s \dot{q}_r - K_i q_r + G_f H + K_s \dot{q}_d + K_i q_d + \ddot{q}_d. \tag{9}$$

Thus, the joint coordinates q have to follow the virtual demand q_r in the ideal case.

2.3 Computation of Compliance Levels

The reference model cannot be arbitrarily determined and it needs to bespoke, suitably adjusted to the context object handled by the robot fingers, in particular when considering the steady state force equilibrium. For this, let us consider the following mass-spring damper system:

$$\ddot{q}_e + \frac{K_{ss}}{m_v}\dot{q}_e + \frac{K_{ii}}{m_v}q_e = \frac{1}{m_v}f \tag{10}$$

where m_v is a virtual mass of the spring, K_{ss} is a virtual damping constant and K_{ii} is a virtual spring constant. By equating equation (10) with equation (8) the following equations are obtained.

$$\frac{K_{ss}}{m_v} = 2\zeta\omega_n = K_s; \quad \frac{K_{ii}}{m_v} = \omega_n{}^2 = K_i; \quad G_f H = \frac{1}{m_v}f \tag{11}$$

where $G_f = \frac{1}{m_v}$, $H = f$, ω_n is the natural frequency and ζ is the damping ratio coefficient. The target is now to determine K_{ss}, K_{ii} and m_v via suitable practical tests and design requirements for compliance and transient behavior. First the actual (constant) demand q_d is given, while q in steady state needs to satisfy the constraint:

$$K_{ii}q_e = f \tag{12}$$

In this case, an initial experiment can be carried out which determines the measurement signal f when the robot finger interacts with a particular object and guarantees safe object handling. Moreover, we may assume that ζ and ω_n are given to establish a suitable transient behavior. Thus, equation (12) and the first two equations in (11) allow now to compute K_{ss} and $\frac{1}{m_v}$ or K_s, K_i and G_f respectively.

2.4 Experimental Setup

The experimental setup for the Berul fingers is shown in Fig. 1. DSPACE DS1006 system for real-time implementation with the BERUL fingers, allow to use MATLAB/Simulink models for fast controller prototyping.

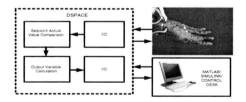

Fig. 1. Experimental setup for the BERUL fingers

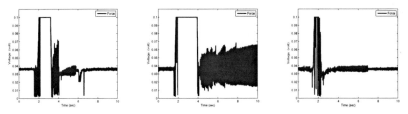

(a) Applied force of a 0.4 (b) Applied Force of a 0.4 (c) Applied force of a 0.4
psi   psi   psi

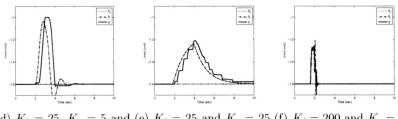

(d) $K_i = 25$, $K_s = 5$ and (e) $K_i = 25$ and $K_s = 25$ (f) $K_i = 200$ and $K_s = 25$
$G_f = 1.67$ and $G_f = 1.67$ and $G_f = 1.67$

Fig. 2. Stiffness level for thumb finger

3   Compliance Level for Different Stiffness

For brevity, only the analysis of different compliance characteristics for the
thumb finger are discussed here. It is noted that from (8) follows $\frac{q_c(s)}{H(s)} = \frac{G_f}{s^2 + K_s s + K_i}$
where $K_s = 2\zeta\omega_n$ and $K_i = \omega_n{}^2$. The scalars ζ and ω_n are damping ratio and
natural frequency respectively. Thus, different K_s and K_i are computed in order
to observe compliance levels.

The compliance model reference behavior is experimentally tested by exerting
a calibrated force of the same amplitude to be sensed by the ISMC algorithm.
In Fig. 2, compliance control results are provided, q_d is the original demand,
q_r is the demand calculated from the virtual reference model (9) and q is the
measured (scaled) angular change of the lowest thumb link. Note that all other
thumb joints are connected to the first joint via a push rod, which links all the
joint angle in an almost linear manner. The results clearly show that our design
for a reference model is effective in creating active compliance. This can be seen
in Fig. 2(e) where by increasing K_s (i.e. increasing the damping coefficient)
the compliance controller becomes sluggish (amplitude of the reference model
demand q_r is about -0.4 rad). On the other hand by increasing K_i (i.e. increasing
the spring coefficient), the compliance becomes more stiff and fast, as seen in
Fig. 2(f) (amplitude of q_r is about -0.3 rad).

4   Results for Grasping Objects

The effectiveness of the ISMC is further tested with different objects namely a *hard* spongy ball (Fig. 3(a)) and a *soft* balloon (Fig. 3(b)). The single tactile pressure sensors (SPTS) which are mounted on the fingers in particular for the thumb, index and ring fingers is also depicted in the above mentioned figures.

The compliant control performance is shown in Fig.4(a) and Fig. 4(b) for the spongy ball and the balloon respectively. It is clearly seen that the compliance results are satisfactorily achieved for different objects (i.e. q follow closely the virtual demand q_r). The reference model is designed so that both objects can be grasped and held up effectively. It is noted that the spongy ball requires higher stiffness, for which $K_i = 20$ while for the balloon $K_i = 15$.

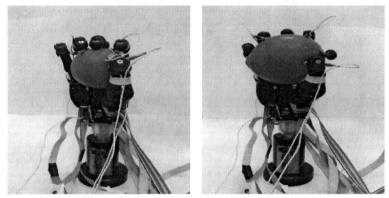

(a) Spongy Ball   (b) Balloon

Fig. 3. Tested objects for practical compliance

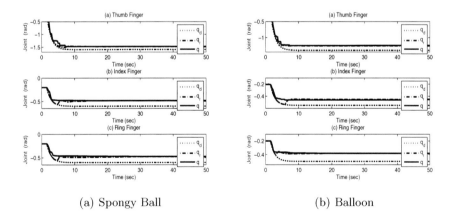

(a) Spongy Ball   (b) Balloon

Fig. 4. Tested objects for practical compliance

5   Conclusions and Future Work

In this paper, we propose a novel approach for active compliant control via Integral Sliding Mode Control (ISMC). Tactile pressure sensors are mounted on the BERUL fingers to measure the interaction force. The effectiveness of the compliance controller when grasping different objects has been successfully demonstrated. It has been shown that levels of compliance can be designed based on the investigated object. Apart from introducing a mass-spring damper reference model, the ISMC is renowned to be robust towards nonlinearity and uncertainties. This is in particular eliminating the friction and stiction emanated from BERUL fingers and touched objects. Future work will consider the automatic design of reference models via experimental tests.

References

[1] Herrmann, G., Melhuish, C.: Towards safety in human robot interaction. International Journal of Social Robotics 2, 217–219 (2010)

[2] Yoshikawa, T.: Multifingered robot hands: Control for grasping and manipulation. Annual Reviews in Control 34(2), 199–208 (2010)

[3] Sisbot, E., Marin-Urias, L., Broqure, X., Sidobre, D., Alami, R.: Synthesizing robot motions adapted to human presence. International Journal of Social Robotics 2, 329–343 (2010)

[4] Wang, W., Loh, R.N., Gu, E.Y.: Passive compliance versus active compliance in robot-based automated assembly systems. Industrial Robot: An International Journal 25(1), 48–57 (1998)

[5] Cutkosky, M.R.: Robotic Grasping and Fine Manipulation. Kluwer Academic Publishers, Norwell (1985)

[6] Johnson, K.L.: Contact Mechanics. Cambridge University Press, Cambridge (1985)

[7] Johnson, K.L.: Contact Problems in the Classical Theory of Elasticity. Alphen aan den Rijn, The Netherlands, Sijthoff and Noordhoff, Netherland (1980)

[8] Shimoga, K., Goldenberg, A.: Soft robotic fingertips. The International Journal of Robotics Research 15(4), 320–334 (1996)

[9] Biagiotti, L., Melchiorri, C., Tiezzi, P., Vassura, G.: Modelling and identification of soft pads for robotic hands. In: 2005 IEEE/RSJ International Conference on Intelligent Robots and Systems (IROS 2005), pp. 2786–2791 (2005)

[10] Liu, H., Hirzinger, G.: Cartesian impedance control for the dlr hand. In: Proceedings of the 1999 IEEE/RSJ International Conference on Intelligent Robots and Systems, IROS 1999 (1999)

[11] Kugi, A., Ott, C., Albu-Schaffer, A., Hirzinger, G.: On the passivity-based impedance control of flexible joint robots. IEEE Transactions on Robotics 24(2), 416–429 (2008)

[12] Albu-Schaffer, A., Ott, C., Hirzinger, G.: A unified passivity-based control framework for position, torque and impedance control of flexible joint robots. The International Journal of Robotics Research 26(1), 23–39 (2007)

[13] Chen, Z., Lii, N., Wimboeck, T., Fan, S., Jin, M., Borst, C., Liu, H.: Experimental study on impedance control for the five-finger dexterous robot hand dlr-hit ii. In: 2010 IEEE/RSJ International Conference on Intelligent Robots and Systems (IROS), pp. 5867–5874 (2010)

[14] Okada, M., Nakamura, Y., Hoshino, S.: Design of active/passive hybrid compliance in the frequency domain-shaping dynamic compliance of humanoid shoulder mechanism. In: Proceedings of the IEEE International Conference on Robotics and Automation, ICRA 2000 (2000)

[15] Montana, D.J.: The Kinematics of Contact and Grasp. The International Journal of Robotics Research 7(3), 17–32 (1988)

[16] Kobayashi, K., Yoshikawa, T.: Controllability of Under-Actuated Planar Manipulators with One Unactuated Joint. The International Journal of Robotics Research 21(5-6), 555–561 (2002)

[17] Jalani, J., Herrmann, G., Melhuish, C.: Robust trajectory following for underactuated robot fingers. In: UKACC International Conference on Control 2010, pp. 495–500 (September 2010)

[18] Canudas De Wit, C., Ge, S.: Adaptive friction compensation for systems with generalized velocity/position friction dependency. In: Proceedings of the 36th IEEE Conference on Decision and Control, vol. 3, pp. 2465–2470 (December 1997)

[19] Ge, S., Lee, T., Ren, S.: Adaptive friction compensation of servo mechanisms. In: Proceedings of the 1999 IEEE International Conference on Control Applications, vol. 2, pp. 1175–1180 (1999)

[20] Shi, J., Liu, H., Bajcinca, N.: Robust control of robotic manipulators based on integral sliding mode. International Journal of Control 81, 1537–1548 (2008)

[21] Yokoyama, M., Kim, G.-N., Tsuchiya, M.: Integral Sliding Mode Control with Anti-windup Compensation and Its Application to a Power Assist System. Journal of Vibration and Control 16, 503–512 (2010)

[22] Defoort, M., Floquet, T., Kokosy, A., Perruquetti, W.: Integral sliding mode control for trajectory tracking of a unicycle type mobile robot. Integr. Comput.-Aided Eng. 13(3), 277–288 (2006)

[23] Eker, I., Akinal, S.: Sliding mode control with integral augmented sliding surface: design and experimental application to an electromechanical system. Electrical Engineering (Archiv fur Elektrotechnik) 90(3), 189–197 (2008)

[24] Chang, J.-L.: Dynamic output integral sliding-mode control with disturbance attenuation. IEEE Transactions on Automatic Control 54, 2653–2658 (2009)

[25] Jalani, J., Herrmann, G., Melhuish, C.: Concept for robust compliance control of robot fingers. In: Proceeding of 11th Conference Towards Autonomous Robotic Systems, pp. 97–102 (May 2010)

Facial Expression Generation Using Fuzzy Integral for Robotic Heads

Bum-Soo Yoo, Se-Hyoung Cho, and Jong-Hwan Kim

Department of Electrical Engineering, KAIST, 355 Gwahangno, Yuseong-gu,
Daejeon, Republic of Korea
{bsyoo,shcho,johkim}@rit.kaist.ac.kr
http://rit.kaist.ac.kr

Abstract. This paper proposes a generation method of facial expressions using fuzzy measure and fuzzy integral for robotic heads. Human's emotion state can be represented by a fuzzy measure which can effectively deal with ambiguity. Because facial expressions are usually ambiguous such that it is difficult to discern emotions and assign a sharp boundary to each emotion. In this method, users can adjust the personality of robot by assignign fuzzy measure to every set of emotions. The partial evaluation values of the current emotion state are obtained from a difference between the ideal basic emotion states and the current emotion state. The Choquet integral of the partial evaluation values with respect to the fuzzy measure is calculated to decide which emotion should occur. The effectiveness of the proposed method is demonstrated through computer simulations and experiments with a robotic head with 19 degrees of freedom, developed in RIT Lab., KAIST.

Keywords: Robotic head, facial expression, fuzzy integral, fuzzy measure.

1 Introduction

Robots have been steadily coming into our daily life. They need interaction capability to provide humans with better service. A desired approach to make natural human-robot interaction (HRI) is learning from human-human interaction (HHI) [1]. In HHI, facial expressions play an important role. According to the psychologist Mehrabian, 55% of information is delivered by non-verbal communications [2],[3]. As facial expressions are essential components in non-verbal communication, they are also important in HRI.

The robotic head Kismet generated facial expressions using a three dimension emotional space and an interpolation [4]. It could produce natural facial expression transitions. However, it could not produce facial expressions dynamically and the interpolation was done by assuming that facial expressions were changing linearly. Besides Kismet, many researches have been progressed to generate natural facial expressions for robotic heads. J.-W. Park used linear dynamic affect-expression model to generate facial expressions for mascot type robotic

T.-H.S. Li et al. (Eds.): FIRA 2011, CCIS 212, pp. 58–65, 2011.

heads[5]. Y. Matsui used recurrent network to consider the time in generating facial expressions[6]. T. B. Bui applied fuzzy rule based system to generate facial expressions,[7]. However, these methods did not consider the personality of the robot. Considering the various facial expressions according to people,even in the same emotion state, personality of the robot should be reflected in generating facial expressions.

In this paper, a generation method of facial expressions for robotic heads using the fuzzy integral and fuzzy measure is proposed. Fuzzy measure is assigned to every sets of emotions according to the personality of the robot. The partial evaluation values of the current emotion state are obtained from a difference between the ideal basic emotion states and the current emotion state. The Choquet integral of the partial evaluation values with respect to the fuzzy measure is calculated to decide which emotion will occur with a certain degree. The effectiveness of the proposed method is demonstrated through computer simulations and experiments with a robotic head with 19 degrees of freedom, developed in RIT Lab., KAIST.

In Section 2, a robotic head is introduced. In Section 3, a method of generating facial expressions using the fuzzy integral and fuzzy measure is proposed for robotic heads. In Section 4, computer simulations and experimental results are presented. Finally, concluding remarks follow in Section 5.

2 Robotic Head

Robotic head's width, height, length and weight are 16 cm, 26 cm, 25 cm and 2.85 kg, respectively. It contains a stereo camera which can receive three dimension images. Totally, 15 servo motors generate facial expressions and 4 DC motors control the movements of neck. Table 1 shows the movements of each part, which is designed based on the facial action coding system [8]. Digital signal processors (DSP) and electrically programmable logic devices (EPLD) control a robotic head. Servo motors are controlled every 20 ms and DC motors are controlled every 5 ms.

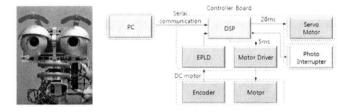

Fig. 1. Robotic head and control diagram

Table 1. Movements of each part

Parts	Degrees of freedom	Movements
Jaw	3	Up and down, left and right, forward and backward
Upper lip, Lower lip	2, 2	Stretch of left and right ends
Eyes	2	Up and down, left and right
Left eye brow, Right eye brow	2, 2	Up and down, left and right
Left eyelid, Right eyelid	1, 1	Up and down
Neck	4	Yaw, pitch, roll axes One more for pitch axis

3　Facial Expression Generation

3.1　Fuzzy Measure and Fuzzy Integral

In this paper, fuzzy measure is used to express the personality of the robot and the Choquet fuzzy integral is used to generate facial expressions. Fuzzy measure means an importance of each set. Among various fuzzy measures, Sugeno λ-fuzzy measure is used to reflect the personality of the robot. Fuzzy measure is defined in the following [9].

Definition 1: A fuzzy measure on the set X of symbols is a set function g : $P(X) \to [0,1]$ satisfying the following axioms:
i) $g(\varnothing) = 0, g(X) = 1$;
ii) $A \subset B \subset X$ implies $g(A) \leq g(B)$.
　　The Sugeno λ-fuzzy measure satisfies the following [10]:

$$g(A \cup B) = g(A) + g(B) + \lambda g(A)g(B). \tag{1}$$

　　Note that users can set a proper λ value, $-1 \leq \lambda \leq \infty$, to adjust the relationship among emotions. If λ is smaller than zero, it means two symbols are in a positive correlation. If λ is greater than zero, it means two symbols are in a negative correlation.
　　The choquet integral is used to generate facial expressions, which is defined in the following [11].

Definition 2: Let h be a mapping from finite set X to $[0,1]$. For $x_i \in X, i = 1, 2, \ldots, n$, assume $h(x_i) \leq h(x_{i+1})$ and $E_i = \{x_i, x_{i+1}, \ldots, x_n\}$. The Choquet fuzzy integral of h over X with respect to the fuzzy measure g is define as

$$\int_X h \circ g = \sum_{i=1}^{n} (h(x_i) - h(x_{i-1}))g(E_i), \ h(x_0) = 0. \tag{2}$$

In (2), fuzzy measure g should be calculated for all the power sets of X. Partial evaluation values h are derived from a difference between the ideal basic emotion states and the current emotion state.

Table 2. Relative importance among emotions

	Anger	Disgust	Fear	Happiness	Sadness	Surprise
Anger	1.0000	0.8333	0.7000	0.5000	0.8000	0.5556
Disgust	1.2000	1.0000	0.8000	0.5556	0.9000	0.6250
Fear	1.4286	1.2500	1.0000	0.6667	0.8000	0.7142
Happiness	2.0000	1.8000	1.5000	1.0000	1.2000	2.0000
Sadness	1.2500	1.1111	1.2500	1.2000	1.0000	0.8333
Surprise	1.8000	1.6000	1.4000	0.5000	1.2000	1.0000

3.2 Weight Assignment

According to the Ekman's research, the number of human's basic emotions is six, i.e. happiness as a positive emotion, surprise as a neutral emotion and anger, disgust fear and sadness as negative emotions [12]. These six parameters are used to represent the robots emotion states and each parameter represent each emotion's degree. It is a difficult task to assign all the fuzzy measures satisfying the axioms of fuzzy measure and to design user's desired personality by hand. In this paper, λ and ϕ_s transformation are used to identify fuzzy measures [13].

Every two pairs should be compared to assign the relative importance in Table 2. Each number means the relative importance of emotion in the row compared to emotion in the column. Therefore, diagonal terms should be one and users should assign the upper triangular values of Table 2. Then lower triangular part will be filled automatically with the reciprocal numbers of the upper triangular values. The normalized values of each row's summation are weights on each emotion. For example, if users want to make a positive personality robot, happiness should more important than other emotions. From Table 2, weight on each emotion is $\{anger, disgust, fear, happiness, sadness, surprise\} = \{0.113, 0.126, 0.151, 0.245, 0.171, 0.193\}$.

3.3 Fuzzy Measure Identification

Based on the weights on each emotion, a fuzzy measure on a set A is identified as follows:

$$\mu_\lambda(A) = \phi_{\lambda+1}\left(\sum_{i \in A} u_i\right) = \frac{\lambda^{u_1+u_2+...+u_n} - 1}{\lambda} \tag{3}$$

where ϕ is the ϕ_s transformation, u_i is the weight of i-th symbol and λ is the interaction degree.

3.4 Partial Evaluation

The current emotion state of the robot is used for calculating the partial evaluation values. Let us denote the current emotion state by a vector $X = [x_1; x_2; \ldots; x_6]^T$, where $x_i, i = 1, 2, \ldots, 6$ is between 0 and 1. The six parameters are used to represent each emotion, i.e. $x_1, x_2, \ldots, x_6$ represents anger, disgust, fear, happiness, sadness, and surprise, respectively.

The partial evaluation values are produced from a difference of the current emotion state vector and the ideal emotion state vectors. Let $S_i = \{s_{i1}; s_{i2}; \ldots; s_{i6}\}^T$, $i = 1, 2, \ldots, 6$ denote the ideal emotion state vector. The ideal emotion state vector is defined as a state vector when a certain emotion is fully occurred. The ideal emotion state vectors parameter $s_{ij}, i, j = 1; 2; \ldots; 6$ is calculated as follows:

$$s_{ij} = \begin{cases} 0 \text{ if } i \neq j \\ 1 \text{ if } i = j. \end{cases} \tag{4}$$

For example, anger is fully occurred when a robot's emotion state vector is $S_1 = \{1; 0; 0; 0; 0; 0\}^T$. $E = \{E_1; E_2; \ldots; E_6\}$ is a distance matrix from the current emotion state vector to the ideal emotion state vectors, which is calculated as follows:

$$\begin{pmatrix} E_1 \\ E_2 \\ \vdots \\ E_5 \\ E_6 \end{pmatrix} = \begin{pmatrix} I - |S_1 - X| \\ I - |S_2 - X| \\ \vdots \\ I - |S_5 - X| \\ I - |S_6 - X| \end{pmatrix} \tag{5}$$

where $E_i = \{e_{i1}; e_{i2}; \ldots; e_{i6}\}^T$ is a size six vector and $I = \{1; 1; 1; 1; 1; 1\}^T$. Each e_{ij} has a different importance in the evaluation of the i-th emotion. For example, when anger expression is generated, anger parameter is more important than other parameters. Therefore, e_{ii} should be considered more than other parameters in E_i. e_{ii}^2 is multiplied to E_i to get partial evaluation vector H_i such that the partial evaluation matrix $H = \{H_1; H_2; \ldots; H_6\}$ is calculated as follows:

$$\begin{pmatrix} H_1 \\ H_2 \\ \vdots \\ H_5 \\ H_6 \end{pmatrix} = \begin{pmatrix} e_{11}^2 \times E_1 \\ e_{22}^2 \times E_2 \\ \vdots \\ e_{55}^2 \times E_5 \\ e_{66}^2 \times E_6 \end{pmatrix} \tag{6}$$

where $H_i = \{h_{i1}; h_{i2}; \ldots; h_{i6}\}^T$ is a size six vector. Each H_i, the partial evaluation value h in (2), is integrated along with the fuzzy measure g from (3) through the Choquet integral. Each H_i are globally evaluated and the biggest emotion is selected to generate facial expressions.

4   Computer Simulation and Experiment

4.1   Simulation

The transition between two similar emotions was simulated. Initial emotion state vector was set as $X = \{Anger, Disgust, Fear, Surprise, Sadness, Surprise\} = \{0.45, x_2, 0.15, 0.08, 0.16, 0.05\}$ and the disgust parameter x_2 was increased from zero to one.

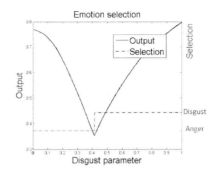

Fig. 2. Transition from anger to disgust

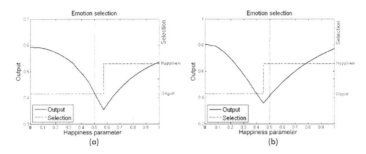

Fig. 3. Comparison between positive personality and negative personality. (a) Positive robot (b) Negative robot

.

In Fig. 2, X, Y1 and Y2-axes represent the disgust parameter, the normalized motor output value and the selected emotion, respectively. When the disgust parameter was zero, anger was selected. As the disgust parameter value was increased, anger was still selected however the magnitude of output decreased because the disgust parameter moved far from zero to one. In addition, the partial evaluation value in (6) is a function of e_{ii}^2 so that the output decreased like second-order equations. The transition occurred when the disgust parameter was around 0.35. After the transition, disgust was selected and its magnitude became larger when the disgust parameter increased.

Comparison between two personalities was simulated. A robot with positive propensity was tuned to have the highest weight on the happiness and robot with negative propensity was tuned to have the highest weight on the disgust. Initial emotion state vector was set as $X = \{Anger, Disgust, Fear, Surprise, Sadness, Surprise\} = \{0.1, 0.5, 0.15, x_4, 0.16, 0.05\}$ and the happiness parameter x_4 was increased from zero to one.

In Fig. 3, X, Y1 and Y2-axes represent the happiness parameter, the output value by a blue line and the selected emotion by a red line, respectively. Black vertical line represents 0.5 which was the same with the disgust parameter's initial value. Fig. 3(a) shows the output of a robot with negative propensity and Fig. 3(b) shows the output of a robot with positive propensity. A positive robot changed to happiness facial expressions in a lower degree of happiness parameter than a negative robot. The difference of emotion selection was caused by the fuzzy measure.

4.2   Experiment

The transition between two opposite emotions was tested. When the disgust parameter was dominant, the happiness parameter increased to make happiness a dominant emotion. Fig. 4 shows the transition from disgust emotion to happiness emotion.

Facial expressions were changed from Fig. 4(a) to Fig. 4(d). The transition had a little discontinuity because at the transition point, facial movements were generated suddenly due to change of selected emotion. However, facial expression transitions between opposite emotions including a transition from disgust to happiness do not usually happen in daily life.

The transition from anger to disgust was tested. Two emotions are so similar that they often appear together and some people are confused in distinguishing them. Facial expressions were changed from Fig. 5(a) to Fig. 5(d). It showed a continuous facial expression transition and did not have any discontinuity as in Fig. 4. Both facial expressions are similar and these facial expression transitions are happened frequently in daily life.

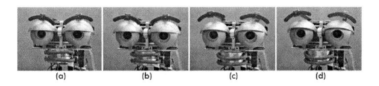

Fig. 4. Transition from disgust to happiness

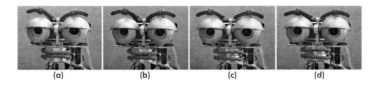

Fig. 5. Transition from anger to disgust

5   Conclusion

In this paper, facial expression generation using the fuzzy measure and fuzzy integral was proposed and applied to a robotic head. The proposed method used the Sugeno λ-fuzzy measure to reflect a robot's personality. The partial evaluation values of the current emotion state were obtained from a difference between the ideal basic emotion states and the current emotion state. The partial evaluation values were integrated along with the fuzzy measure through the Choquet integral to generate facial expressions. The proposed method was demonstrated by the experiments with a robotic head and it could generate natural facial expressions.

References

1. Muhl, C., Nagai, Y.: Does Disturbance Discourage People from Communication with a Robot? In: 16th IEEE International Conference on Robot & Human Interactive Communication, Jeju, Korea, pp. 1137–1142 (2007)
2. Mehrabian, A.: Nonverbal Communication. Aldine-Atherton (1972)
3. Mehrabian, A.: Communication without words. Psychology Today 2(4) (1968)
4. Breazeal, C.L.: Designing Sociable Robots. MIT Press, Cambridge (2002)
5. Park, J.-W., Kim, W.-H., Lee, W.-H., Kim, W.-H., Chung, M.-J.: Lifelike Facial Expression of Mascot-Type Robot based on Emotional Boundaries. In: IEEE Int. Conf. Robotics and Biomimetics, Guilin, China, pp. 830–835 (2009)
6. Matsui, Y., Kanoh, M., Kato, S., Nakamura, T., Itoh, H.: Evaluating A Model for Generating Interactive Facial Expressions using Simple Recurrent Network. In: IEEE Int. Conf. Syst. Man Cybern., Texas, USA, pp. 1639–1644 (2009)
7. Bui, T., Heylen, D., Poel, M., Nijholt, A.: Generation of Facial Expressions from Emotion Using a Fuzzy Rule Based System. In: 14th Australian Joint Conf. Artificial Intelligence, Adelaide, Australia (2001)
8. Ekman, P., Friesen, W.: Facial Action Coding System: A Technique for the Measurement of Facial Movement. Consulting Psychologists Press, Palo Alto (1978)
9. Sugeno, M.: Theory of fuzzy integrals and its applications. Ph.D. dissertation, Tokyo Institute of Technology (1974)
10. Kim, J.-H., Han, J.-H., Kim, Y.-H., Choi, S.-H., Kim, E.-S.: Preference-based Solution Selection Algorithm for Evolutionary Multiobjective Optimization. IEEE Transactions on Evolutionary Computations (December 2010) (accepted)
11. Grabisch, M., Nguyen, H.-T., Walker, E.-A.: Fundamentals of uncertainty Calculi, with Applications to Fuzzy Inference. Kluwer, Dordrecht (1995)
12. Ekman, P., Friesen, W.: Unmasking the face: A Guide to Recognizing Emotions From Facial Expressions. Prentice Hall, Englewood Cliffs (1975)
13. Takahagi, E.: On identification methods of λ-fuzzy measures using weights and λ. Journal of Japan Society for Fuzzy Theory and Systems 12(5), 665–676 (2000)

Implementations and Controls of a 3-DOF Parallel Link Joint Module

Po-Chun Chia[1], Cheng-Wei Dong[2], and Chung-Hsien Kuo[1]

[1] National Taiwan University of Science and Technology,
106 Taipei, Taiwan, R.O.C.
{d9903105,chkuo}@mail.ntust.edu.tw
[2] Chang Gung University,
333 Tao-Yuan, Taiwan, R.O.C.
m9625015@stmail.cgu.edu.tw

Abstract. In this paper, a parallel link joint module (PLJM) is developed to achieve three degree-of-freedom (DOF) joint motions with respect to a fixed center of rotation. The proposed 3-DOF spatial PLJM is configured with three linear screw actuators, a base platform, a movable platform, a central bar, and seven ball joints. Due to similar motion characteristics, the PLJM can be used to construct 3-DOF joints of humanoid robots such as hip, ankle, and shoulder joints. Meanwhile, the motions of neck, waist, and wrists can be also constructed using the PLJM. In addition to parallel kinematics mechanism features, the linear screw actuator based PLJM also provides power saving benefits. That is, the PLJM only consumes energy when the PLJM works. As a consequence, humanoid robots constructed using the PLJM may provide longer service time when compared to rotary based joints. Structurally, a PLJM is desired as the connection of a 3-DOF joint and a follower limb; therefore, the PLJM may simplify the mechanical structure of humanoid robots. In addition to the PLJM design, the kinematics, trajectory controls, and interactive sensor integrations are implemented in this paper. Several interesting experiments are demonstrated to verify our approaches.

Keywords: parallel robots, path planning for manipulators, mechanism design.

1 Introduction

Parallel kinematics [1]-[3] are popularly used in robotic manipulators. Advantages of parallel mechanisms are low inertia, high rigidity, simple structure, zero-accumulation error, high static loading capacity, etc. Parallel link based manipulators are widely applied to machine tools [4], medical procedures [5], humanoid robots [6], gait simulations [7], etc. Because the manipulation path of a parallel mechanism may perform similar spatial motion characteristics with human beings, parallel mechanisms are practical to construct the joint of humanoid robots such as the neck [8], ankle [9], and shoulder [10]. At the same time, the parallel mechanism can be also used to develop the walking chair [11].

By exploring the body skeleton structure and the joint motion characteristics of human beings, 3-DOF spatial motions appear in most of joints. The hip, ankle, and

T.-H.S. Li et al. (Eds.): FIRA 2011, CCIS 212, pp. 66–77, 2011.

shoulder are ball joints. In addition, although the neck, waist, and wrist are not fitted into ball joints, they may be approximated as 3-DOF spatial joints for simplicity when their motions are investigated.

In general, biped humanoid robots are developed to perform similar morphological structures and motion characteristics with human beings. From the viewpoints of a human's skeleton structure, degree of freedoms and range of motions are crucial design issues for a humanoid robot. Motions of a humanoid robot are generated via driving the joints. Therefore, the joint design hugely affects the motion performance of humanoid robots.

Joints of most humanoid robots are developed using rotary motor and gear trains. Three rotary gear motors are organized in three independent orthogonal axes to form a conventional 3-DOF spatial joint. Nevertheless, the gear backlash, mechanism complexity and size of such joints are major design problems for humanoid robots. In order to reduce the backlash as well as the weight and size of gear trains, expensive harmonic drive gear trains are used [12]. Meanwhile, the inertia, loading capacity, and accumulated errors are other concerns of rotary gear motor configurations.

From the biomechanics viewpoints of human beings, ball joint structures are composed of a bone frame with a ball socket and a diaphysis with a ball head, as shown in Fig. 1. Relative motions appear between the bone frame and diaphysis.

For human beings, joints are driven in terms of parallel tendons, and concentrations of muscle skeleton fibers (tendons) result in the motions of the corresponding joint. Therefore, the 3-DOF joint of a humanoid robot can be developed based on this concept. This idea also appeared in [13]. Based on this design concept, relative motions between the bone frame and diaphysis can only be achieved in a parallel driven manner. As a consequence, the 3-DOF joint can be driven by controlling the connected parallel links.

In this paper, a PLJM is developed to simulate 3-DOF joint motions with respect to a fixed center of rotation. The proposed 3-DOF spatial PLJM is configured with three linear screw actuators, a base platform, a movable platform, a central bar, and seven ball joints. Several literatures used pneumatic sources to drive parallel links such as [13]-[14], and they may result in the portable and mobile problems. In this study, linear screw actuators are used to provide convenient power sources, simplified wiring, and accurate control performance.

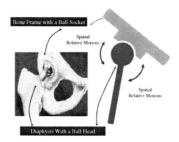

Fig. 1. 3-DOF joint structure of a biomechanics investigation

On the other hand, power consumptions are also another issue for biped humanoid robots. The power consumption represents a huge difference between wheeled and biped robots. Discarding the number of motors, the wheel robot consumes little power when the robot stops. However, the humanoid robot continuously consumes energy when the robot stands still if rotary gear motors are used. The reason is that continuous power outputs of rotary gear motors on the joints have to overcome the gravity of body weights. If the rotary gear motor stops its output power, the humanoid robot will fall down.

Contrarily, the linear screw actuator is a non-reversible mechanism. That means the motor can drive the screw so that the output translational motion can be desired; nevertheless, the translational motion is impossible to result in the rotations of the screw. Therefore, the humanoid robot can stand still without any power outputs using the linear screw actuators. This feature may save large amount of energy when the humanoid robot does not move frequently. This concept is beneficial to the applications of exhibitions and guards that need long time services and infrequent movements. More specially, depletions of powers and failures of motors will not result in the danger of suddenly falling down.

Although the linear screw actuator based PLJM performs better power saving solutions for humanoid robots when compared to conventional rotary gear motor solutions, slow responses of screw actuators and non-reversible motions are the drawbacks. Therefore, the linear screw actuator based PLJM may be appropriate for the humanoid robots with slow walk speed. On the other hand, current feedback approaches can be further developed for the linear screw actuator to ensure safety. If the safety of the PLJM can be guaranteed in the future, this study would be applied to the prostheses.

Finally, this paper is organized as follows: section 2 describes the mechanical design; section 3 elaborates the kinematics and trajectory control; section 4 represents the controller implementations and interactive sensor integrations; section 5 demonstrates the experiments and results; and the conclusions and future works are summarized in section 6.

2 Mechanical Design

In this paper, a parallel link joint module (PLJM) is developed to simulate 3-DOF spatial joint motions of human beings. The proposed 3-DOF spatial PLJM is configured with three linear screw actuators, a base platform, a movable platform, a central bar, and seven ball joints. A central bar is vertically fixed on the base platform at its proximal side. The distal side of the central bar connects a ball joint, and this ball joint further connects to a movable platform via a ball socket.

Three linear screw actuators are used to form controllable parallel links. The DC motor, gear trains, linear screw, and magnetic type encoders are installed inside the linear screw actuator. Two ball joints are connected to two sides of the linear screw actuator. Each ball joint of a linear screw actuator connects to the base platform and the movable platform, respectively. The photos of a linear screw actuator with two ball joints and an assembled PLJM is shown in Fig. 2.

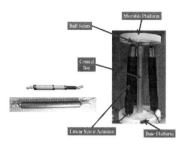

Fig. 2. Linear screw actuator and assembled PLJM module

The mobility of the PLJM is also discussed. This spatial linkage consists of eight links, three screw translation joints, and seven ball joints. By referring to the Kutzbach criterion [15], the mobility of the PLJM is 6 where 3 of them can be treated as idle degrees of freedom. That means the symmetric rotations of linear screw actuator will not affect relative motions of the movable platform and the based platform. Therefore, three linear screw actuations are capable of manipulating the movable platform of a PLJM with a desired orientation within the mechanism reachable workspace.

The link parameters of this PLJM are summarized in table I. The movable platform of a PLJM is initially placed in parallel to the based frame, and the elements of an initial posture vector are all zeros in three rotary coordinates. The based and movable platforms are made using engineering plastic. The overall height of a PLJM is 315 mm, and the weight is 4.4 Kg. The radius of central bar is 15 mm.

Such PLJM modules can be used to construct the hip, ankle, shoulder, neck, waist, and wrist joints of a biped humanoid robot. An example is shown in Fig. 3. This robot is configured with 14 joint modules (totally, 34 DOFs), where 10 of them use the PLJM modules. This structure is quite simple because most of the components are assembled using the proposed PLJM modules.

3 Kinematics and Trajectory Control

Based on the mechanism design of Fig. 2, the inverse kinematics of a PLJM is evaluated for the link length control purposes. The kinematic diagram is shown in Fig. 4. The posture vector is defined as $\theta = [\theta_x, \theta_y, \theta_z]^T$, and it represents the relative orientation of a movable platform with respect to (w.r.t.) a base platform. O is the center (origin) of a base platform, and e is the center (origin) of a movable platform. A_1, A_2, and A_3 are coordinates of ball joints on a base platform with respect to origin O. b_1, b_2, and b_3 are coordinates of ball joints on a movable platform with respect to origin e. Zb_1, Zb_2, and Zb_3 are coordinates of ball joints on a movable platform with respect to origin e for a zero posture vector ($\theta = [0, 0, 0]^T$). B_1, B_2, and B_3 are coordinates of ball joints on a movable platform with respect to origin O. L_1, L_2, and L_3 are lengths of linear screw actuators. $T_h = [0, 0, h]^T$, and it indicates a fixed coordinate translation between the base and movable platforms.

Table 1. Link Parameters

Symbol	Quantity	Description
Φ	120 mm	Diameter of base frame
ϕ	100 mm	Diameter of movable frame
h	300 mm	Length of central bar
R	80 mm	Distance of ball joint to base frame center
r	50 mm	Distance of ball joint to movable frame center
ZL_1	$\sqrt{h^2 + (R-r)^2}$	Initial length of linear screw actuator # 1
ZL_2	$\sqrt{h^2 + (R-r)^2}$	Initial length of linear screw actuator # 2
ZL_3	$\sqrt{h^2 + (R-r)^2}$	Initial length of linear screw actuator # 3

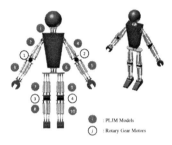

Fig. 3. Constructing a humanoid robot using 10 PLJM modules

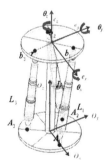

Fig. 4. Kinematic diagram of a PLJM module

At the beginning of operations, initial coordinates of each frame are defined as shown in (1) – (6).

$$A_1 = [R\ cos(0),\ R\ sin(0),\ 0]^T \tag{1}$$

$$A_2 = [R\ cos(2\pi/3),\ R\ sin(2\pi/3),\ 0]^T \tag{2}$$

$$A_3 = [R\ cos(4\pi/3),\ R\ sin(4\pi/3),\ 0]^T \tag{3}$$

$$Zb_1 = [r\ cos(0),\ r\ sin(0),\ 0]^T \tag{4}$$

$$Zb_2 = [r\ cos(2\pi/3),\ r\ sin(2\pi/3),\ 0]^T \tag{5}$$

$$Zb_3 = [r\ cos(4\pi/3),\ r\ sin(4\pi/3),\ 0]^T \tag{6}$$

By introducing the Eular-Angle representations, any posture change of the movable platform ($\theta = [\theta_x, \theta_y, \theta_z]^T$) will result in a new set of coordinates of b_1, b_2, and b_3. In a practical application, the control of a PLJM module needs to find the posture angles of θ_x, θ_y, and θ_z. In this study, the problem of defining the posture angles of θ_x, θ_y, and θ_z are developed by knowing the normal vector (P) and the roll angle (θ_z) of a movable platform.

In general, the roll angle (θ_z) of a movable platform is desired by users for different manipulation purposes. The rotations with respect to X and Y axes may result in the transitions of normal vectors. θ_x defines the coordinate rotations with respect to X axis, and θ_y defines the coordinate rotations with respect to Y axis. Different rotation sequences of θ_x and θ_y may result in different postures of the movable platform. Therefore, the first task is to define the rotation sequence for a given normal vector, $P = [P_x, P_y, P_z]^T$ so that the axis of a movable platform (e_z) is colinear with vector P.

Initially, a point P in the e coordinate system is to transform as e_z, and then these angles are used to find posture vectors. As consequence, the transformation matrix can be found. Detailed procedures are elaborated as the following steps.

1. Rotate P w.r.t. X axis to the X-Z plane with the angle θ_{yz}.

$$\theta_{yz} = \tan^{-1}(P_y / P_z) \tag{7}$$

2. Transform P to X-Z plane w.r.t. X axis with θ_{yz} as $P_1 = (P_{1x}, 0, P_{1z})$. Note that the y- component must be zero because point P_1 is on the X-Z plane. The coordinates transform matrix ($TmpR_x$) [16] is used for this transformation, as shown in (8) – (9). Where s indicates $sin(\cdot)$; and c indicates $cos(\cdot)$

$$TmpR_x = \begin{bmatrix} 1 & 0 & 0 \\ 0 & c\theta_{yz} & -s\theta_{yz} \\ 0 & s\theta_{yz} & c\theta_{yz} \end{bmatrix} \tag{8}$$

$$P_1 = TmpR_x \times P \tag{9}$$

3. Rotate P_1 w.r.t. Y axis to the Z axis with the angle θ_{xz}. The transformed coordinate will be on the Z-axis.

$$\theta_{yz} = -\tan^{-1}(P_x / P_z) \tag{10}$$

4. Based on the knowledge of θ_{yz}, and θ_{xz}, the x- and y- components of a posture vector can be defined in terms of inverting the signs of θ_{yz} and θ_{xz}, respectively. Equations of (11) – (13) show the results.

$$\theta_x = -\theta_{yz} \tag{11}$$

$$\theta_y = -\theta_{xz} \tag{12}$$

5. Based on the posture vector of $[\theta_x, \theta_y, \theta_z]^T$, the Z-axis with a roll angle can be properly transformed to a desired normal vector (P) via inverting the transformation sequences of step 1 and step 3. A resulting transformation matrix for this posture vector is indicated in (13).

$$R_{trans} = R_x(\theta_x)R_y(\theta_y)R_z(\theta_z)$$

$$= \begin{bmatrix} 1 & 0 & 0 \\ 0 & c\theta_x & -s\theta_x \\ 0 & s\theta_x & c\theta_x \end{bmatrix} \begin{bmatrix} c\theta_y & 0 & s\theta_y \\ 0 & 1 & 0 \\ -s\theta_y & 0 & c\theta_y \end{bmatrix} \begin{bmatrix} c\theta_z & -s\theta_z & 0 \\ s\theta_z & c\theta_z & 0 \\ 0 & 0 & 1 \end{bmatrix} \qquad (13)$$

Based on these steps, the coordinates of three ball joints with respect to a given normal vector and a roll angle can be obtained by multiplying the R_{trans} with their initial positions. At the same time, these coordinates w.r.t. O coordinate system can also be calculated. These equations are shown in (14) – (15).

$$b_i = R_{trans} \times Zb_i, \text{ where } i = 1 \text{ to } 3. \qquad (14)$$

$$B_i = bi + T_h, \text{ where } i = 1 \text{ to } 3. \qquad (15)$$

Finally, the inverse kinematics can be solved by finding the length of three linear screw actuators using the Euclidean distance, as shown in (16).

$$L_i = \| Ai - Bi \|, \text{ where } i = 1 \text{ to } 3. \qquad (16)$$

On the other hand, to avoid intersections of parallel links, three parallel links' vectors and a central bar vector are evaluated, as shown in (17) – (18).

$$V_0 = \overline{e - O} \qquad (17)$$

$$V_i = \overline{A_i - B_i}, \text{ where } i = 1 \text{ to } 3. \qquad (18)$$

The minimum distance (d_{ij}) of any two link vectors i and j ($i, j = 0$ to 3; $i \neq j$) can be evaluated as (19). Where A_i can be referred to Fig. 4, and A_0 indicates the origin of base frame (O).

$$d_{i,j} = \left| \overline{A_i A_j} \bullet \frac{\overline{V_i} \times \overline{V_j}}{|\overline{V_i} \times \overline{V_j}|} \right|, \text{ where } i, j = 0 \text{ to } 3. \qquad (19)$$

The intersection of a PLJM is defined as the minimum distance (d_{ij}) of any two links is less than the sum their link cylinders' radius. Usually, a safety factor is also used to overcome the control and structure uncertainties. If the intersection is happened, the desired posture is not reachable.

In summary, the trajectory control for a PLJM module can be divided as the following procedures:

1. Define the sequence of spatial normal vectors and roll angles of the movable platform according to manipulation purposes.
2. Apply these spatial normal vectors and roll angles to calculate the corresponding posture vectors and coordinate transformation matrices.
3. Use the coordinate transformation matrix to obtain ball joint coordinates of the movable platform.
4. Evaluate the reachability of a desired movable platform posture using the minimum distance between any two parallel links.

5. Calculate the lengths of three linear screw actuators between the based and movable platforms if the desired posture is reachable. At the same time, the length should be within the stroke of the linear screw actuator. These lengths will be used to control the corresponding actuators so that the movable platform may reach a desired orientation.

4 Implementations and Interactive Sensor Integrations

4.1 Controller Implementations

In addition to the mechanism design, PLJM fabrication, kinematics, and trajectory control, a PLJM control system is also developed. The control system is composed of three closed-loop motor controllers. Each closed-loop motor controller is implemented using a DSPic30F2010 [17] based microcontroller, and it is responsible of UART communications, motor encoder counting, closed loop proportional–integral–derivative (PID) position servo control algorithm, and generating pulse-width modulation (PWM) outputs, as shown in Fig. 5.

Three DC motor controllers are integrated to control a 3-DOF PLJM module via half-duplex (universal asynchronous receiver/ transmitter) UART communications. Each DC motor controller is identified using an ID (1 to 3) for individual communication purposes, as shown in Fig. 6. On the other hand, in order to verify the performance of trajectory control, a laser pointer (with control ID: 0) is used. The laser pointer is mounted on the center of a movable platform, and its laser beam indicates the normal direction of this movable platform. The laser beam may project on a parallel plate which is parallel to the base frame. By capturing continuous exposure in the same frame of a camera, a continuous trajectory can be recorded to evaluate the performance of trajectory controls.

It is noted that the singularity of the PLJM can be simply resolved using slightly control delay of three actuators when the current posture of a PLJM is in a singularity, such as the initial zero posture vector. This operation may resolve the uncertainty of movements (non-unique solution of forward kinematics) when the PLJM is in a singularity situation.

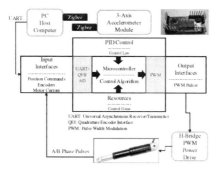

Fig. 5. Closed loop motor control architecture

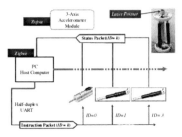

Fig. 6. Control network topology of PLJM manipulations

4.2 Interactive Sensor Integrations

The PLJM may simulation human's 3-DOF joint motions. In order to demonstrate its 3-DOF joint motion capability, a tri-axis accelerometer (type: Hitachi H48C [18]) module combining with a wireless sensor node (BAT mote [19]) is attached on human's head and wrist to measure the joint motions of the neck and wrist, as shown in Fig. 7. The human's joint motions are further transmitted to the PC via the Zigbee communications, as shown in Fig. 5 – 6. By using wireless sensor network, it is more convenient for joint motion measurements.

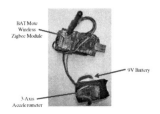

Fig. 7. Tri-axis accelerometer module with Zigbee wireless sensor node

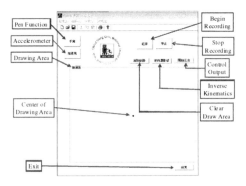

Fig. 8. Graphical user interface for PLJM control

Finally, a graphical user interface is implemented using the Microsoft Visual C++ [20] for the collections of pen-pad data and human's joint motion signal as well as trajectory controls, as shown in Fig. 8.

5 Experiments and Results

5.1 Pen Drawing Path Experiments

Two pen drawing experiments are done via writing a word and drawing a star using a commercial pen-pad. The pen trajectories will be recorded, and then converted to normal vectors of the movable platform. The posture vectors are further calculated to generate the lengths of linear screw actuators. This program may automatically divide the whole trajectory into individual path segments, so that the output of the laser pointer can be desired. A camera is continuously exposing to form an entire path on the same frame. Fig. 9 shows the experimental results. Two charts indicate the control length (in motor shaft encoder pulses) of three linear screw actuators for two experiments. These results verify the proposed trajectory control approaches. Nevertheless, the trajectory control performance can be further improved in terms of enhancing assemble accuracy of the PLJM and reducing backlashes of linear screws.

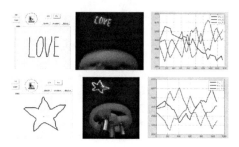

Fig. 9. Trajectory controls of pen-pad inputs for a 'LOVE' word and a star

5.2 Human's 3-DOF Joint Motion Following

The human's 3-DOF joint motion following capacity is also evaluated. The first experiment is done by mounting the tri-accelerometer on the user's wrist. The PLJM moves its movable platform following the wrist's postures. Fig. 10 shows the experimental results of five different wrist postures. In addition, another experiment is done by mounting the tri-accelerometer on the user's head. The PLJM also moves its movable platform following the head's postures. Fig. 11 shows the experimental results of five different head postures. Hence, these experiments successfully demonstrate that the PLJM may follow 3-DOF joint motions of the wrist and neck.

Fig. 10. Trajectory control of following the wrist joint motions

Fig. 11. Trajectory control of following the neck joint motions

6 Conclusion

In this paper, a PLJM module is proposed. The PLJM modules behaves similar joint motion characteristic with human's 3-DOF joints. An experimental PLJM prototype is produced in our laboratory. The kinematics, trajectory control, and hardware implementations are all discussed in this paper. In the future, a full size humanoid robot will be developed by using the proposed PLJM modules to verify the feasibility of a PLJM based humanoid robot.

Acknowledgments. This work is supported by National Science Council under Grant number 95-2221-E-011-226-MY3.

References

1. Yang, G., Chen, I.M., Chen, W., Lin, W.: Kinematic Design of a Six-DOF Parallel-kinematics Machine with Decoupled-motion Architecture. IEEE Transactions on Robotics 20(5), 876–887 (2004)
2. Seo, T.W., Kang, D.S., Kim, H.S., Kim, J.: Dual Servo Control of a High-tilt 3-DOF Microparallel Positioning Platform. IEEE/ASME Transactions on Mechatronics 14(5), 616–625 (2009)
3. Pierrot, F., Nabat, V., Company, O., Krut, S., Poignet, P.: Optimal Design of a 4-DOF Parallel Manipulator: from Academia to Industry. IEEE Transactions on Robotics 25(2), 213–224 (2009)

4. Wang, L., Wu, J., Wang, J., You, Z.: An Experimental Study of a Redundantly Actuated Parallel Manipulator for a 5-DOF Hybrid Machine Tool. IEEE/ASME Transactions on Mechatronics 14(1), 72–81 (2009)
5. Li, Y., Xu, Q.: Design and Development of a Medical Parallel Robot for Cardiopulmonary Resuscitation. IEEE/ASME Transactions on Mechatronics 12(3), 265–273 (2007)
6. Alfayad, S., Ouezdou, F.B., Namoun, F., Bruneau, O., Henaff, P.: Three DOF Hybrid Mechanism for Humanoid Robotic Application: Modeling, Design and Realization. In: IEEE/RSJ International Conference on Intelligent Robots and Systems, pp. 4955–4961 (2009)
7. Aubin, P.M., Cowley, M.S., Ledoux, W.R.: Gait Simulation via a 6-DOF Parallel Robot with Iterative Learning Control. IEEE Transactions on Biomedical Engineering 55(3), 1237–1240 (2008)
8. Sabater, J.M., Garcia, N., Perez, C., Azorin, J.M., Saltaren, R.J., Yime, E.: Design and Analysis of a Spherical Humanoid Neck Using Screw Theory. In: IEEE/RAS-EMBS International Conference on Biomedical Robotics and Biomechatronics, pp. 1166–1171 (2006)
9. Liu, G., Gao, J., Yue, H., Zhang, X., Lu, G.: Design and Kinematics Analysis of Parallel Robots for Ankle Rehabilitation. In: IEEE/RSJ International Conference on Intelligent Robots and Systems, pp. 253–258 (2006)
10. Lenarcic, J., Stanisic, M.: A Humanoid Shoulder Complex and the Humeral Pointing Kinematics. IEEE Transactions on Robotics and Automation 19(3), 499–506 (2003)
11. Marco, C., Giuseppe, C.: A New Leg Design with Parallel Mechanism Architecture. In: IEEE/ASME International Conference on Advanced Intelligent Mechatronics, pp. 1447–1452 (2009)
12. Kaneko, K., Harada, K., Kanehiro, F., Miyamori, G., Akachi, K.: Humanoid Robot HRP-3. In: IEEE/RSJ International Conference on Intelligent Robots and Systems, pp. 2471–2478 (2008)
13. Zhu, X., Tao, G., Yao, B., Cao, J.: Adaptive Robust Posture Control of Parallel Manipulator Driven by Pneumatic Muscles with Redundancy. IEEE/ASME Transactions on Mechatronics 13(4), 441–450 (2008)
14. Takuma, T., Hayashi, S., Hosoda, K.: 3D Bipedal Robot with Tunable Leg Compliance Mechanism for Multi-modal Locomotion. In: IEEE/RSJ International Conference on Intelligent Robots and Systems, pp. 1097–1102 (2008)
15. Gogu, G.: Chebychev–Grübler–Kutzbach's Criterion for Mobility Calculation of Multi-loop Mechanisms Revisited via Theory of Linear Transformations. European Journal of Mechanics - A/ Solids 24(3), 427–441 (2005)
16. Kuo, C.H., Lee, M.Y., Huang, C.C., Hung, K.F., Chiu, Y.S.: Development of 3D Navigation System for Retained Auricular Prosthesis Application. Journal of Medical and Biological Engineering 23(3), 149–158 (2003)
17. Microchip DSPic30F2010 datasheet,
http://ww1.microchip.com/downloads/en/DeviceDoc/70118e.pdf
18. Hitachi H48C tri-axis accelerometer datasheet,
http://www.parallax.com/dl/docs/prod/acc/HitachiH48C3AxisAccelerometer.pdf
19. BAT mote product information,
http://www.bandwavetech.com/en/en_index.htm
20. Microsoft visual C++ product information, http://www.microsoft.com/

Motion Recognition in Wearable Sensor System Using an Ensemble Artificial Neuro-Molecular System

Si-Jung Ryu and Jong-Hwan Kim

Department of Electrical Engineering, KAIST, 355 Gwahangno, Yuseong-gu,
Daejeon, Republic of Korea
{sjryu,johkim}@rit.kaist.ac.kr
http://rit.kaist.ac.kr

Abstract. This paper proposes an ensemble artificial neuro-molecular system for motion recognition for a wearable sensor system with 3-axis accelerometers. Human motions can be distinguished through classification algorithms for the wearable sensor system of two 3-axis accelerometers attached to both forearms. Raw data from the accelerometers are pre-processed and forwarded to the classification algorithm designed using the proposed ensemble artificial neuro-molecular(ANM) system. The ANM system is a kind of bio-inspired algorithm like neural network. It is composed of many artificial neurons that are linked together according to a specific network architecture. For comparison purpose, other algorithms such as artificial neuro-molecular system, artificial neural networks support vector machine, k-nearest neighbor algorithm and k-means clustering, are tested. In experiments, eight kinds of motions are randomly selected in a daily life to test the performance of the proposed system and to compare its performance with that of existing algorithms.

Keywords: Artificial neoro-molecular system (ANM), Motion recognition, Wearable sensor system, Ensemble network.

1 Introduction

A rapid development in computer technology has imposed a new computing environment. A computer is combined with an intelligent human-friendly interface and will appear a new computing environment. This concept is represented by a ubiquitous computing. Most of all, wearable computing leads a next generation computing based on ubiquitous computing. Wearable computing includes not only intelligent computer that is worn to the body, but also just sensors attached to the body. Nowadays, wearable computing technology has been used in a variety of fields including sports, medical care, the game, etc. There are also many researches about wearable computing, and exist two big issues about wearable computing. One is a new computer environment that is combined with human-friendly interface. Another is a wearable health care system that can help patients and senior man at long distance.

T.-H.S. Li et al. (Eds.): FIRA 2011, CCIS 212, pp. 78–85, 2011.

This paper proposes an ensemble artificial neuro-molecular (ANM) system for motion recognition for a wearable sensor system with 3-axis accelerometers considering a human-friendly interface. Many researches related to wearable computing based on human-friendly interface have been conducted. In particular, human motion recognition using sensor systems is widely used. Hardware platforms based on accelerometers are most popular. Other kinds of sensors such as gyro sensors, pressure sensors and cameras can also be used for the hardware platforms [1]. The sensors are attached to various locations such as forearms, wrists, head, waist, legs, etc [2]. As for the classification, algorithms such as classifier including neural network, support vector machine, k-means clustering, k-nearest neighbor, etc., can be used [3]–[7].

For comparison purpose, the performance of the proposed ensemble ANM system is compared with that of those algorithms. To demonstrate the effectiveness of the proposed system, experiments are carried out for eight kinds of motions that are randomly selected in a daily life. Also, its performance is compared with that of existing algorithms including neural network, k-nearest neighbor, support vector machine, and k-means clustering.

The rest of this paper is organized as follows. In Section 2, the ensemble artificial neuro-molecular system is proposed. Section 3 describes the wearable sensor system The experimental results are discussed in Section 4 and concluding remarks follow in Section 5.

2   Ensemble Artificial Neuro-Molecular System

2.1   Artificial Neuro-Molecular System (ANM)

Artificial neuro-molecular system (ANM) is a biologically motivated system that captures the biological structure-function relationships, and it possesses several features that facilitate evolutionary learning [9].

Overall Structure: ANM mainly consists of four neurons, receptor neurons, cytoskeletal neurons, reference neurons, and effector neurons. Receptor neurons receive input data from outside and transform into an internal signal. Cytoskeletal neurons receive a signal from receptor neurons and fire an effector neuron. Then finally, an effector neuron determines a class of input data. Reference neurons play a role to supervise cytoskeletal neurons. The overall strucutre of ANM is depicted in Fig. 1.

Cytoskeletal Neuron: A cytoskeletal neuron is composed of 8×8 sized site array. Each site can have one of the three-types of components (C1, C2, or C3) or none. Each site can also have MAP, readout enzyme, and readin enzyme. A MAP links two neighboring components of different types together. Specific combinations of cytoskeletal signals will activate a readout enzyme, which causes the neuron to fire. A readin enzyme converts an external signal into a cytoskeletal signal. For the details of signal flow in a cytoskeletal neuron, the reader is referred to [8], [9].

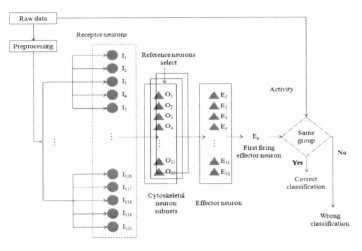

Fig. 1. Overall structure of ANM system

Reference Neuron: Two-layered reference neurons supervise cytoskeletal neurons. High-level reference neuron chooses a bunch of low-level reference neurons, while each low-level reference neuron chooses cytoskeletal neurons. Only selected cytoskeletal neurons receive a signal from receptor neurons. The structure of the reference neurons are described in Fig. 2.

Two-level Evolutionary Learning: In the ANM system, evolutionary learning is used as a learning method. It is progressed in two level, cytoskeletal neuron level and reference neuron level as follows.

I. Evolutionary learning at cytoskeletal neuron level

1) Calculate the fitness value of each subnet.
2) According to the fitness value, copy the best three subnets to other subnets.
3) Mutate some of neurons in other subnets.

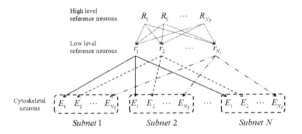

Fig. 2. Structure of reference neuron

II. Evolutionary learning at reference neuron level

1) Calculate the fitness value of each high level reference neuron.
2) According to the fitness value, copy the best reference neuron to other reference neurons.
3) Mutate some of neurons in other subnets.

2.2 Ensemble ANM System

A single ANM system has poor performance if the number of classfication categories increases. In this case, the performance can be improved by constructing ensemble network. There are several kinds of combination method which have been frequently used: voting, weighted sum, Bayesian, etc. The results from each independent network are integrated by a combination method which yields a new single result. In this paper, th voting combination method is used. The key of the voting combination is a majority vote. The final result is decided as the one that most of individuals choose. In this paper, ensemble network is constructed, which is composed of eight individual networks, as shown in Fig. 3.

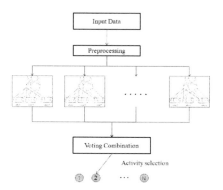

Fig. 3. Ensemble ANM system

3   Wearable Sensor System

3.1   Hardware Structure

Hardware platform of the wearable sensor system mainly consists of accelerometers and a communication module. Freescale MMA7260Q triaxial accelerometer is selected as a measurement device, which has four different sensitivities, 1.5g, 2.0g, 4.0g, 6.0g. In this paper, 1.5g is used and two accelerometers, each of which needs 3.3V for battery power, are attached to forearms, respectively, and a communication module to a waist, as shown in Fig. Fig. 4. As for the micro-controller, ATmega128 is used, which needs 5V for battery power. Considering both of the accelerometer and micro-controller, a 5V dry cell type battery is used so that a battery is directly connected to an ATmega128 and connected

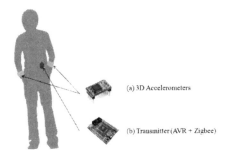

Fig. 4. Locations of sensors and a communication module

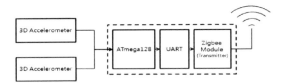

Fig. 5. Internal structure of the wearable sensor system

to the two accelerometers through a diode that drops electric pressure by 1.7V. For communication, Zigbee module is used, which is connected to UART in ATmega128 board. The internal structure of the wearable sensor system is shown in Fig. 5.

3.2 Data Collection

The output of the accelerometer is analog data, so ATmega128 converts it into 10-bit digital data. Also, acceleration data from the two accelerometers is collected for each sampling time of 100ms for 3 seconds. In other words, acceleration data of which length is 3 seconds will be an input data of classification algorithm. In this case, there are two drawbacks: over-fitting problem and heavy computational complexity. Thus, a series of pre-processing steps is needed to solve the problems. The steps are executed by data sampling, removal gravity acceleration and quantization in order.

3.3 Pre-processing

Data Sampling: Raw acceleration data is obtained at 50Hz, but the mean of every five acceleration data is used to reduce the complexity of the algorithm.

Removal Gravity Acceleration: Gravity acceleration is included to an acceleration data from the acceleromter. Because we are interested in a trajectory of the acceleration data, an initial reference value is subtracted from the accelration data. After this pre-processing, all initial values are set to 0 g.

Quantization: The ADC values are divided into five intervals and transformed into 5 bit data. The quantization rule used is as follows:

$$
\text{Quantized\ \ data} = \begin{cases}
00001 & \text{if} & -300.0 < \text{actual value} < -180.0 \\
00010 & \text{if} & -180.0 < \text{actual value} < -60.0 \\
00100 & \text{if} & -60.0 < \text{actual value} < 60.0 \\
01000 & \text{if} & 60.0 < \text{actual value} < 180.0 \\
10000 & \text{if} & 180.0 < \text{actual value} < 300.0
\end{cases} .
$$

4   Experiments

4.1   Experimental Setup

We randomly selected eight kinds of motions in a daily life. Each motion is defined in Table 1. A training set was obtained by repeating ten times for each motion, and a test set was obtained by repeating five times for each motion. One motion data was composed of six accelerometer values for 3 seconds.

Table 1. Motions

Motion No	Motion
1	Halt
2	Swing two arms
3	Shake two arms back and forth
4	Stretch two arms forward
5	Put the hands behind the head
6	Greeting
7	Raise a right arm
8	Intend to hit something with left arm

4.2   Experimental Results

In the ANM system, mutatin rate for neurons, the number of receptor neuron, the number of subnets, and cytoskeletal neurons in a subnet were set to 0.1, 120, 8, and 32, repectively. The termination condition was 800 generations. We used multi-layered perceptron (MLP) trained with backpropagation for neural networks. Three hidden layers were used, and each hidden layer contains 5 neurons. In the case of k-NN, we classifed motions based on closest training examples in the 8-dimensional euclidean space. Finally, a 3rd order polynomial kernel function was used for SVM because input data was non-linear.

Table 2a, 2b shows that k-NN and SVM performed best. However, there are drawbacks in k-NN and SVM. First, the performance of SVM is heavily influenced by kernel functions which transform support vector. Actually, SVM had the worst performance when inappropriate kernel functions were used. Second, the performance of k-NN is dependent on k and types of motions. In other words, k-NN might have poor performance for the classification problem of complex motions. Artificial neuro-molecular (ANM) system has a similar structure

Table 2. Accuracy for training and test data

<table>
<tr><td colspan="2">(a) Training data</td><td colspan="2">(b) Test data</td></tr>
<tr><td>Classification Algorithm</td><td>Training Data Accuracy</td><td>Classification Algorithm</td><td>Test Data Accuracy</td></tr>
<tr><td>ANM</td><td>61.25%</td><td>ANM</td><td>53.75%</td></tr>
<tr><td>ANN</td><td>68.75%</td><td>ANN</td><td>46.00%</td></tr>
<tr><td>k-NN</td><td>-</td><td>k-NN</td><td>95.00%</td></tr>
<tr><td>SVM</td><td>96.25%</td><td>SVM</td><td>85.00%</td></tr>
<tr><td>K-Means Clustering</td><td>83.75%</td><td>K-Means Clustering</td><td>47.50%</td></tr>
<tr><td>Ensemble ANM</td><td>91.25%</td><td>Ensemble ANM</td><td>72.50%</td></tr>
<tr><td>Ensemble ANN</td><td>86.75%</td><td>Ensemble ANN</td><td>75.00%</td></tr>
</table>

as artificial neural network. Both algorithms have randomness and we do not know how to classify data internally. In additions, both algorithms can classify the data without specific pre-processing like a kernel. It means that performances of classification is less affected by the input data. Therefore, it is reasonable to compare the performances of ANN and ANM system. In actual practice, ANN and ANM system have a similar performance and ANM system's performance is slightly better than ANN in accuracy for training data. It is also shown that the proposed ensemble ANM network improves classification accuracy.

5   Conclusions

This paper proposed a novel ensemble artificial neuro-molecular (ANM) system and applied it to motion classification in a wearable sensor system. The system classified human motions when she/he wearing the system did some different motions. Artificial neuro-molecular system employed two-level evolutionary learning. As the performance of a single ANM system was poor, the ensemble network of ANM systems was developed. As a result, classification accuracy was improved as much as general classification algorithms' accuracy. To test the performance of the proposed ensemble ANM system, eight motions were randomly selected in a dailtly life. The ANM system gave an accurate classification result for the eight motions. For comparison purpose, neural network, k-nearest neighbor, support vector machine and k-means clustering were tested for the same motions. As a result of the comparison, the performance of the proposed ensemble ANM system was similar as that of other algorithms. Other sensors such as gyro sensors, ECG sensors, image sensors as well as accelerometers could improve the performance, which is left for future work.

Acknowledgments. This research was supported by the MKE (The Ministry of Knowledge Economy), Korea, under the National Robotics Research Center for Robot Intelligence Technology support program supervised by the NIPA (National IT Industry Promotion Agency) (NIPA-2010-N02100128).

References

1. Krause, A., Siewiorek, D.P., Smailagic, A., Farringdon, J.: Unsupervised, Dynamic Identification of Physiological and Activity Context in Wearable Computing. In: 7th IEEE International Symposium on Wearable Computers, pp. 88–97. IEEE Press, New York (2003)
2. Kern, N., Schiele, B., Schmidt, A.: Multi-sensor activity context detection for wearable computing. In: Aarts, E., Collier, R.W., van Loenen, E., de Ruyter, B. (eds.) EUSAI 2003. LNCS, vol. 2875, pp. 220–232. Springer, Heidelberg (2003)
3. Ravi, N., Dandelkar, N., Mysore, P., Littman, M.L.: Activity Recognition from Accelerometer Data. In: 17th Conference on Innovative Applications of Artificial Intelligence, pp. 1541–1546. AAAI Press, Pittsburgh (2005)
4. Bao, L., Intille, S.S.: Activity Recognition from User-Annotated Acceleration Data. In: Ferscha, A., Mattern, F. (eds.) PERVASIVE 2004. LNCS, vol. 3001, pp. 1–17. Springer, Heidelberg (2004)
5. DeVaul, R.W., Dunn, S.: Real-Time Motion Classification for Wearable Computing Applications. Technical report, MIT Media Lab (2001)
6. Huynh, T., Schiele, B.: Analyzing Features for Activity Recognition. In: Proceedings of the 2005 Joint Conference on Smart Objects and Ambient Intelligence, Grenoble, pp. 159–163 (2005)
7. Long, X., Yin, N., Aarts, R.M.: Single-Accelerometer-Based Daily Physical Activity Classification. In: 31st Annual International Conference of the IEEE Engineering in Medicine and Biology Society, pp. 6107–6110. IEEE Press, Minneapolis (2009)
8. Lee, W.-C., Chen, J.-C., Hsu, C.-C.: Health analysis from the trend variability of patients' vital signs with an artificial neuromolecular system. In: IEEE World Congress on Computational Intelligence, pp. 2794–2799. IEEE Press, Barcelona (2010)
9. Chen, J.-C., Conrad, M.: Pattern categorization and generalization with a virtual neuromolecular architecture. Neural Networks 10(1), 111–123 (1997)
10. Cho, S.-B., Won, H.-H.: Cancer classification using ensemble of neural networks with multiple significant gene subset. Applied Intelligence 26(3), 243–250 (2007)

A SoPC-Based Surveillance System

Yuan-Pao Hsu[*], Hsiao-Chun Miao, and Sheng-Han Huang

Department of Computer Science and Information Engineering,
National Formosa University,
Yunlin, Twiwan 632, R.O.C.
hsuyp@nfu.edu.tw

Abstract. This article focuses on how to use simple image processing techniques to realize a dynamic object detecting and tracking surveillance system on a SoPC. We try to mount a camera on a two-dimensional rotation machinery so as to dynamically search the environment by controlling the rotation of this machinery. In detection mode, the system rotates the machinery along a predefined path to capture images with fixed time interval and compare the images with their corresponding previously recorded reference images for determining if any intrusion objects appear. Once an intrusion object is detected, the system switches to tracking mode. In tracking mode, successive images are compared to find the most possible area in the image where the object locates. The color which occupies biggest region in this possible area in the image is finally recognized as the feature of the intrusion object. The resulting system has functions including intrusion detecting, object tracking, warning message sending, and internet remote watching and all these functions have been experimentally proven that they works well on the SoPC system simultaneously.

Keywords: SoPC, FPGA, image processing, object tracking.

1 Introduction

Recently, surveillance systems with image recording functions become vital devices both in private and public places. The reason is that, according to the recorded images, the police have resolved many serious robberies and criminal cases which are difficult in the past and the responsibilities of traffic accident cases have been clarified in terms of the saved images as well. Therefore, the government has put the construction of image recording systems on top of the list of the public security infrastructure. However, the surveillance system has usually been designed and installed in the way that its camera has only fixed direction that limits its monitor function. Consequently, it would require many cameras for the system to keep a large area under watching. Such that how to design a surveillance system with low cost and large secure area becoming the purpose of this design.

There are two major approaches on real-time image tracking. The first approach is to track objects according to their features. As study in [1], vehicles were tracked basing on images captured by the traffic monitors; and in [2], in accordance with local binary pattern and skin color, human face could be tracked.

---

[*] Corresponding author.

T.-H.S. Li et al. (Eds.): FIRA 2011, CCIS 212, pp. 86–93, 2011.

The second approach is to distinguish objects from background from images. From [3], a method was proposed to estimate the traffic flow by computing differences of images for the purpose of extracting object edges for vehicles tracking. A block-based motion estimation method was used to tracking a moving object in [4]. The way that subtracts two images to obtain moving object information is often employed in the case when background is steady. Comparing input image with the example image, the different area will be referred to be the moving objects. This method is simple in concept, however when background is not steady it becomes very difficult to separate moving objects from background. Besides, it consumes much computation effort to search the whole image for motion estimation. The drawbacks prevent this method from satisfying applications which require real-time process.

There are more moving object tracking algorithms. In [5], for instance, a particle filter concept was employed for object tracking according to similarity of color density function to predict the object position. A histogram based method was developed to process consecutive images for object tracking in [6]. In study [7], an adaptive block matching algorithm was proposed for tracking, and in [8], multiple objects tacking was examined by combining features extraction and moving properties of objects. These methods emphasize high positioning rate and recognizing rate, but on the opposite side they need complex computation so they usually implemented only by PCs or embedded systems.

The rapid progress of FPGA (Field Programmable Gate Array) technologies makes it possible for many image processing methods being successfully realized on hardware in recent decades. Birla used a FPGA to build an image processing platform involving interfacing of the FPGA to CMOS image sensor and VGA monitor [13]. In research of Zhang et al, the FPGA technology was used for fast median filter and denoising processing of images to help improve the algorithm of image clarity processing [14]. For image tracking, Cho et al proposed multiple objects tracking by using particle filter method and implemented on a FPGA [15]. Other literature about FPGA implementation of image processing algorithms can be seen in [16].

This article combines aforementioned two approaches to detect an object by distinguishing the object from the background, and to track according to the features of the object. By controlling a two-dimensional rotation machinery on which the camera is mounted, the system can track a suspected object and provide the police with saved images of the object for further detecting reference.

The organization of this paper is as follows: functions of each major block of the system are introduced in section 2. Section 3 describes the operation and function of each major circuit including hardware and software design of the image process circuit. The design of system software, interfacing of remote control, accessing of SD card and sounding program are shown in Section 4, whereas Section 5 discusses the design result. Conclusion and future perspective are drawn in Section 6.

2   System Architecture

Fig. 1 depicts the architecture of the proposed system. In the figure, DE2-70 is the SoPC platform. Images captured by CMOS camera are processed by image processing unit to detect and track any intrusion objects. The system controls RC (radio controlled) motors

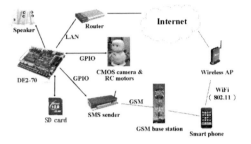

Fig. 1. System architecture

of the two-dimensional machinery to aim at the object for tracking and saves captured images into SD card at the same time. Speaker alarms the sound stored in SD card when suspected object is detected. Meanwhile, the surveillance system includes other functions e.g., sending warning message through the SMS sender, controlling the direction of camera through the internet.

3   System Hardware Design

Hardware design flow of the image processing unit is shown in Fig. 2 and is elucidated as follows.

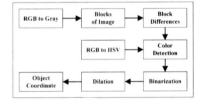

Fig. 2. Image processing unit design flow

3.1   RGB to Gray

Color image is firstly transformed into gray image in order to reduce the computation burden and register usage. The transform equation is as (1).

$$Gray = \frac{R+G+B}{3}. \tag{1}$$

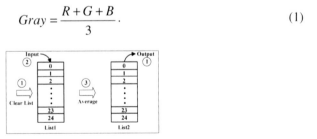

Fig. 3. Operation of partitioning an image into blocks

3.2 Blocks of Image

This step partitions the gray image into 5x5 blocks and the 25 average gray values are recorded. Fig. 3 shows the process of this step:

(1) Lists initialization: clear contents of List1 and List2.
(2) Accumulation of gray value of each block: the image gray value of each block is summed up and stored in List1.
(3) Average value computation: summarized values in List1 are averaged for and saved in List2.

3.3 Block Difference

The background image and present captured image are partitioned into 5x5 blocks and each block's gray value is computed by following the previous mentioned operation. The difference of every two related blocks of the background image and present image is computed to determine if an object intrusion is detected, when the difference is higher than a threshold.

3.4 RGB to HSV

True color RGB mode uses three basic colors, red, green and blue to represent the color of pixels of an image. Each color is encoded by 8 bits so a color value varies ranging from 0 to 255, such that, for single basic color, the color difference can be identified from this value. However, if three basic colors are mixed it may cause the problem that two colors are only slightly different in value but are recognized to be two totally different colors in vision instead. In other words, there is a gap between color and vision. This problem jeopardizes the object identification of the system. We design a converter to convert the image from RGB to HSV format to solve this problem [17].

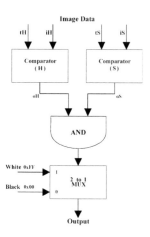

Fig. 4. Circuit of color detection and binarization

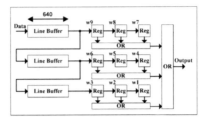

Fig. 5. Image dilation circuit

3.5 Color Detection and Binarization

The circuit for color detection and binarization is demonstrated in Fig. 4. In the circuit, comparators compare input image data to check if the H and S of pixels of the data fall into the range of interest or not by following the rules described in (5) and (6), if so, set the pixels to white; to black otherwise.

$$oH(iH) = \begin{cases} 1 & H_{min} \le iH \le H_{max} \\ 0 & otherwise \end{cases}, \tag{2}$$

$$oS(iS) = \begin{cases} 1 & S_{min} \le iS \le S_{max} \\ 0 & otherwise \end{cases}, \tag{3}$$

3.6 Image Dilation

The purpose of dilating an image is mainly to make up possible flaws of the image caused by noises. Dilation process sets the binary value of a pixel to 1 if any of its proximity pixels have binary value of 1.

The circuit architecture of the dilation unit can be seen in Fig. 5. The OR output will be 1 if any of the eight registers are 1 and this result is written into the register w5 to accomplish the dilation process. The three line buffers each has 640 cells are used to accommodate three image lines of a 640 x 480 image data.

4 System Software Design

System software of the designed system is in charge of three functions: (1) transforming the captured images into BMP format and saving them on the SD card with filenames according to the time and date when the files are saved; (2) alarming the speaker through LINEOUT on the board by reading sound file from SD card; (3) providing real time image on home page and remote control function to control the camera direction through internet. The basic hardware system for handling these software functions is illustrated in Fig. 6. In fact the Nios II processor and other circuits in Fig. 6 are embedded together with the hardware system which runs the image processing task as described in last section on the same FPGA chip.

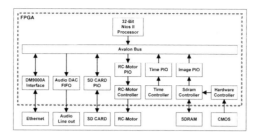

Fig. 6. System architecture of Niso II processor

5   System Integration

The processing speed would be very slow and inefficient if we use software program to implement the entire surveillance system. By using hardware software co-design methodology, the image processing functions as described in section 3 and hardware base system for software execution as described in section 4 are integrated as demonstrated in Fig. 7.

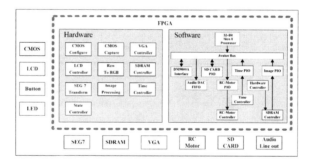

Fig. 7. Integrated architecture of the surveillance system

6   Design Result and Performance

The resultant system is implemented on an Altera Cyclone II 2C35 FPGA which provides 35,000 of logic elements to be used. Table I. lists the FPGA resources usage of the system. It takes 85% of resources of the FPGA to accommodate the accomplished system. The hardware, including image processing and others, use 55,749 LEs of the FPGA, in which image processing circuit consumes 15,360 LEs that is about 22% of the FPGA resources.

The size of an image frame is 800 x 525 for the image processing unit to process. Because the system clock rate is 25MHz, it takes about 16.8ms to process an image frame and it means that the resulting system only needs 16.8ms to identify and locate the suspected object on an image frame.

Table 1. FPGA resources usage of resulting system

FPGA(35,000LEs)						
System			**LEs**	**Total (LEs)**	**HW (LEs)**	**Resource usage**
Hardware	Image Processing	RGB to HSV	1851	15360	55749	22%
		Face Skin Detection & Binarization	67			
		Dilation	24			
		Object Coordinate	141			
		RGB to Gray	79			
		Image Frame	306			
		Diff Frame	1419			
		Image Block	1677			
		Circle Block	3199			
		Diff Block	612			
		Read Color	5907			
		Track Controller	78			
	Other		40389			56%
Software			5187			7%
Total Resources usage						85%

7 Conclusion and Future Work

This paper realizes an object detecting and tracking system on a SoPC. Using the hardware and software co-design methodology we pet together the entire system on the FPGA of a SoPC to achieve the goals of small size, low power consumption, and high speed on handling functions of:

(1) dynamic object detecting and tracking,
(2) scanning path setting,
(3) alarm sounding,
(4) real time image saving,
(5) warning message sending, and
(6) internet remote monitoring.

In the future the designed system can be enhanced by following approaches as follows.

(1) The rotation of the two-dimensional machinery is based on looking up a table causing that when the object is moving too slow or too fast that the machine is unable to lock the object precisely. It is desirable to use more accurate methods to control the rotation of the machine.
(2) At the moment, the system locks an object only by skin color. The detecting and tracking of the system would be more precise if we can add human face identification and recognition methods onto it.

References

1. Robert, K.: Video-based traffic monitoring at day and night vehicle features detection tracking. In: IEEE Conference on Intelligent Transportation Systems, pp. 285–290 (2009)
2. Wang, C.-X., Li, Z.-Y.: A new face tracking algorithm based on local binary pattern and skin color information. In: IEEE Transactions on International Symposium on Computer Science and Computational Technology, vol. 2, pp. 657–660 (2008)
3. Dailey, D.J., Cathey, F.W., Pumrin, S.: An algorithm to estimate mean traffic speed using uncalibrated cameras. IEEE Transactions on Intelligent Transportation Systems, 98–107 (2000)
4. Shin, H.S., Kim, S.M., Kim, J.W., Eom, Hwan, K.: Novel object tracking method using the block-based motion estimation. In: IEEE Transactions on SICE Annual Conference, pp. 2535–2539 (2007)
5. Islam, M.Z., Oh, C.-M., Lee, C.-W.: Real time moving object tracking by particle filter. In: IEEE Transactions on International Symposium on Computer Science and its Applications, pp. 347–352 (2008)
6. Zhao, A.: Robust histogram-based object tracking in image sequences. IEEE Transactions on Digital Image Computing Techniques and Applications, 45–52 (2007)
7. Hu, W.-C.: Adaptive template block-based block matching for object tracking. IEEE Transactions on Intelligent Systems Design and Applications 1, 61–64 (2008)
8. You, W., Jiang, H., Li, Z.-N.: Real-time multiple object tracking in smart environments. In: IEEE Conference on Robotics and Biomimetics, pp. 818–823 (2009)
9. Li, C.: Design of image acquisition and processing based on FPGA. IEEE Transactions on Information Technology and Applications, 113–115 (2009)
10. The RGB Color Space, http://gimp-savvy.com/BOOK/index.html?node50.html
11. The HSV Color Space, http://www.blackice.com/colorspaceHSV.htm
12. Wu, S.-L., Lin, H.-T.: Digital image processing using C, pp. 81–82. Taiwan Chuan Hua Inc. (2008) (in Chinese)
13. Jinghong, D., Yaling, D., Kun, L.: Development of Image Processing System Based on DSP and FPGA. In: IEEE Conference on Electronic Measurement and Instruments, pp. 2-791–2-794 (2007)
14. Shanthi, K.J., Ashok, L.R., Anandu, A.S., Das, B.G.: FPGA Implementation of Image Segmentation Processor. In: IEEE Conference on Emerging Trends in Engineering and Technology, pp. 364–367 (2009)
15. Li, C.X., Huang, Y.P., Han, X.X., Pang, J.: Algorithm of binary image labeling and parameter extracting based on FPGA. In: IEEE Conference on Computer Engineering and Technology, vol. 3, pp. 542–545 (2010)
16. Zhang, M., Li, F., Li, J.Y.: The research of real-time image clarity processing method based on FPGA. In: IEEE Conference on Information Technology in Medicine and Education, pp. 1302–1306 (2009)
17. The HSV Color Space, http://www.blackice.com/colorspaceHSV.htm

Intelligent Motion Control for Four-Wheeled Omnidirectional Mobile Robots Using Ant Colony Optimization

Hsu-Chih Huang

Department of Computer Science and Information Engineering,
Hungkuang University,
34 Chung-Chie Rd, Sha Lu, Taichung, 443, Taiwan, R.O.C.
hchuang@sunrise.hk.edu.tw

Abstract. This paper presents an intelligent motion controller for four-wheeled omnidirectional mobile robots with four independent driving wheels equally spaced at 90 degrees from one another by using ant colony optimization (ACO). The optimal parameters of motion controller are obtained by minimizing the performance index using the metaheuristic ACO algorithm. These optimal parameters are used in the ACO motion controller to obtain better performance for four-wheeled omnidirectional mobile robots to achieve both trajectory tracking and stabilization. Simulation results are conducted to show the effectiveness and merit of the proposed ACO-based intelligent motion controller for four-wheeled omnidirectional mobile robots.

Keywords: ACO, kinematic, mobile robot, optimization.

1 Introduction

Recently, omnidirectional mobile robots have attracted much attention in both academia and industry in the field of robotics. Such robots are superior to those with differential wheels in terms of dexterity and driving ability [1-3]. They are shown to perform various movements difficult or impossible for differential wheeled mobile robots [1-3]. Comparing with several car-like robots [4-6], the type of omnidirectional mobile mechanism has the superior agile capability to move towards any position and to attain any desired orientation. To date, a variety of omnidirectional mobile robots have been proposed in [1-3]. Among these omnidirectional mobile robots, the one with four or more wheels are shown more powerful than that with three wheels [7].

Modeling and control of four-wheeled omnidirectional mobile robots have been investigated by several researchers [7-11]. Byun *et al.* [8] constructed a four-wheeled omnidirectional mobile robot with a variable wheel arrangement mechanism. Purwin *et al.* [9] presented an optimal trajectory planning approach for a four-wheel omnidirectional mobile robot. Shing *et al.* [10] proposed a T-S fuzzy path controller design for a four-wheeled omnidirectional mobile robot. The dynamic model and fuzzy controller for a four-wheeled omnidirectional surveillance robot were presented in [11]. Moreover, Tsai *et al.* [12] presented a dynamic model incorporating frictions and

T.-H.S. Li et al. (Eds.): FIRA 2011, CCIS 212, pp. 94–106, 2011.

dynamic effects to achieve motion control. However, these studies did not cope with the controller parameter optimization problems.

There are many methods to address the optimal problem of mobile robots [13-16]. Among these approaches, ACO proposed by Dorigo [17] has been regarded as an effective metaheuristic algorithm in finding optimal solutions for difficult combinatorial problems [17-19]. In ACO computing, a set of artificial ants search for good solutions for the optimal problems. Each ant constructs a solution by making a sequence of local decisions which are guided by pheromone information and some heuristic information. After a number of ants have constructed solutions, the best ants then update the pheromone information along their path. This algorithm has many advantages, including combing distributed computation, positive feedback and constructive greedy heuristic. However, there has no attempt to using ACO to design an intelligent motion controller for four-wheeled omnidirectional mobile robots.

The objective of this paper is to design an intelligent motion controller based on ACO algorithm for four-wheeled omnidirectional mobile robots to achieve both trajectory tracking and stabilization. With the kinematic model of the omnidirectional mobile robot, the unified control law is proposed to achieve stabilization and trajectory tracking for the mobile robots. Moreover, the controller parameters are then optimized by using the ACO algorithm; thereby improving the controller performance. The rest of this paper is organized as follows. In Section 2, the kinematic control law is proposed to achieve stabilization and trajectory tracking for the four-wheeled omnidirectional mobile robots. Section 3 elaborates the ACO algorithm and its application to controller parameter tuning. Section 4 conducts several simulations to show the effectiveness and merit of the proposed method. Section V concludes this paper.

2 Kinematic Control

This section is devoted to briefly describing the kinematic model of an omnidirectional mobile robot with four independent driving wheels equally spaced at 90 degrees from one another. With the kinematic model, a kinematic controller is proposed to achieve stabilization and trajectory tracking.

2.1 Kinematic Model

Fig. 1 depicts the structure and geometry of the four-wheeled omnidirectional driving configuration with respect to a world frame. Due to structural symmetry, the vehicle has the property that the center of geometry coincides with the center of mass. In what follows describes the kinematic model of this kind of robot, where θ represents the vehicle orientation which is positive in the counterclockwise direction. Note that θ also denotes the angle between the moving frame and the world frame. On the basis of the method proposed by [1], it is easy to obtain the following inverse kinematic model of the four-wheeled omnidirectional mobile platform in the world frame.

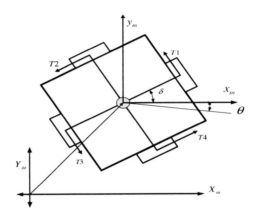

Fig. 1. Structure and geometry of the four-wheeled omnidirectional mobile robot

$$
\upsilon(t) = \begin{bmatrix} \upsilon_1(t) \\ \upsilon_2(t) \\ \upsilon_3(t) \\ \upsilon_4(t) \end{bmatrix} = \begin{bmatrix} r\omega_1(t) \\ r\omega_2(t) \\ r\omega_3(t) \\ r\omega_4(t) \end{bmatrix} = P(\theta(t)) \begin{bmatrix} \dot{x}(t) \\ \dot{y}(t) \\ \dot{\theta}(t) \end{bmatrix} \tag{1}
$$

where

$$
P(\theta(t)) = \begin{bmatrix} -\sin(\delta+\theta) & \cos(\delta+\theta) & L \\ -\cos(\delta+\theta) & -\sin(\delta+\theta) & L \\ \sin(\delta+\theta) & -\cos(\delta+\theta) & L \\ \cos(\delta+\theta) & \sin(\delta+\theta) & L \end{bmatrix} \tag{2}
$$

and $\omega_i(t), i = 1,2,3,4$ respectively denotes the angular velocity of each wheel; r denotes the radius of each wheel; L represents the distance from center of the platform to the center of each wheel. Note that although the matrix $P(\theta(t))$ is singular for any θ, but its left inverse matrix can be found, i.e., $P^{\#}(\theta(t))P(\theta(t)) = I$, and expressed by

$$
P^{\#}(\theta(t)) = \begin{bmatrix} \dfrac{-\sin(\delta+\theta)}{2} & \dfrac{-\cos(\delta+\theta)}{2} & \dfrac{\sin(\delta+\theta)}{2} & \dfrac{\cos(\delta+\theta)}{2} \\ \dfrac{\cos(\delta+\theta)}{2} & \dfrac{-\sin(\delta+\theta)}{2} & \dfrac{-\cos(\delta+\theta)}{2} & \dfrac{\sin(\delta+\theta)}{2} \\ \dfrac{1}{4L} & \dfrac{1}{4L} & \dfrac{1}{4L} & \dfrac{1}{4L} \end{bmatrix} \tag{3}
$$

2.2 Kinematic Control

With the kinematic model in (1), this subsection is devoted to designing two kinematic controllers to achieve point-to-point stabilization and trajectory tracking for the omnidirectional mobile robot in Fig. 1. Furthermore, a unified nonlinear control approach is also presented as below.

2.2.1 Point Stabilization

The control goal of the point stabilization is to find the controlled angular velocity vector $[\omega_1(t) \quad \omega_2(t) \quad \omega_3(t) \quad \omega_4(t)]^T$ to steer the mobile robot from any starting pose $[x_0 \quad y_0 \quad \theta_0]^T$ to any desired destination pose $[x_d \quad y_d \quad \theta_d]^T$. Note that the current pose of the mobile robot is $[x(t) \quad y(t) \quad \theta(t)]^T$. To design the controller, one defines the pose error which is the difference between the present pose and the desired destination pose, that is,

$$
\begin{bmatrix} x_e(t) \\ y_e(t) \\ \theta_e(t) \end{bmatrix} = \begin{bmatrix} x(t) \\ y(t) \\ \theta(t) \end{bmatrix} - \begin{bmatrix} x_d \\ y_d \\ \theta_d \end{bmatrix}
\tag{4}
$$

which gives

$$
\begin{bmatrix} \dot{x}_e(t) \\ \dot{y}_e(t) \\ \dot{\theta}_e(t) \end{bmatrix} = \begin{bmatrix} \dot{x}(t) \\ \dot{y}(t) \\ \dot{\theta}(t) \end{bmatrix} = P^{\#}(\theta(t)) \begin{bmatrix} r\omega_1(t) \\ r\omega_2(t) \\ r\omega_3(t) \\ r\omega_4(t) \end{bmatrix}
\tag{5}
$$

To asymptotically stabilize the system, the following stabilization law is proposed. Note that the matrices K_P and K_I are symmetric and positive definite, i.e., $K_P = diag\{K_{p1}, K_{p2}, K_{p3}\} = K_P^T > 0,\ K_I = diag\{K_{i1}, K_{i2}, K_{i3}\} = K_I^T > 0.$

$$
\begin{bmatrix} \omega_1(t) \\ \omega_2(t) \\ \omega_3(t) \\ \omega_4(t) \end{bmatrix} = \frac{1}{r} P(\theta(t)) \left(-K_P \begin{bmatrix} x_e(t) \\ y_e(t) \\ \theta_e(t) \end{bmatrix} - K_I \begin{bmatrix} \int_0^t x_e(\tau)d\tau \\ \int_0^t y_e(\tau)d\tau \\ \int_0^t \theta_e(\tau)d\tau \end{bmatrix} \right)
\tag{6}
$$

Taking (6) into (5), the dynamics of the closed-loop error system becomes

$$
\begin{bmatrix} \dot{x}_e(t) \\ \dot{y}_e(t) \\ \dot{\theta}_e(t) \end{bmatrix} = -K_P \begin{bmatrix} x_e(t) \\ y_e(t) \\ \theta_e(t) \end{bmatrix} - K_I \begin{bmatrix} \int_0^t x_e(\tau)d\tau \\ \int_0^t y_e(\tau)d\tau \\ \int_0^t \theta_e(\tau)d\tau \end{bmatrix}
\tag{7}
$$

For the asymptotical stability of the closed-loop error system, a radially unbounded Lyapunov function candidate is chosen as follows:

$$
V_1(t) = \frac{1}{2}[x_e(t) \quad y_e(t) \quad \theta_e(t)] \begin{bmatrix} x_e(t) \\ y_e(t) \\ \theta_e(t) \end{bmatrix}
$$

$$
+ \frac{1}{2} \left[\int_0^t x_e(\tau)d\tau \quad \int_0^t y_e(\tau)d\tau \quad \int_0^t \theta_e(\tau)d\tau \right] K_I \begin{bmatrix} \int_0^t x_e(\tau)d\tau \\ \int_0^t y_e(\tau)d\tau \\ \int_0^t \theta_e(\tau)d\tau \end{bmatrix}
\tag{8}
$$

Taking the time derivative of $V_1(t)$, one obtains

$$\dot{V}_1(t)=\begin{bmatrix} x_e(t) & y_e(t) & \theta_e(t) \end{bmatrix}\begin{bmatrix} \dot{x}_e(t) \\ \dot{y}_e(t) \\ \dot{\theta}_e(t) \end{bmatrix}+\begin{bmatrix} \int_0^t x_e(\tau)d\tau & \int_0^t y_e(\tau)d\tau & \int_0^t \theta_e(\tau)d\tau \end{bmatrix}K_I\begin{bmatrix} x_e(t) \\ y_e(t) \\ \theta_e(t) \end{bmatrix}$$

$$=-\begin{bmatrix} x_e(t) & y_e(t) & \theta_e(t) \end{bmatrix}K_p\begin{bmatrix} x_e \\ y_e \\ \theta_e \end{bmatrix}<0 \tag{9}$$

Since $\dot{V}$ is negative semidefinite, Barbalat's lemma implies that $\begin{bmatrix} x_e(t) & y_e(t) & \theta_e(t) \end{bmatrix}^T \rightarrow \begin{bmatrix} 0 & 0 & 0 \end{bmatrix}^T$ as $t\rightarrow\infty$.

2.2.2 Trajectory Tracking

This subsection considers the trajectory tracking problem of the omnidirectional mobile robot. Unlike all nonholonomic conventional mobile robots, the trajectories of the omnidirectional mobile robots can not be generated using their kinematic models, i.e., any smooth and differentiable trajectories for the omnidirectional robots can be arbitrarily planned. Given the smooth and differentiable trajectory $\begin{bmatrix} x_d(t) & y_d(t) & \theta_d(t) \end{bmatrix}^T \in C^1$, one defines the following tracking error vector

$$\begin{bmatrix} x_e(t) \\ y_e(t) \\ \theta_e(t) \end{bmatrix}=\begin{bmatrix} x(t) \\ y(t) \\ \theta(t) \end{bmatrix}-\begin{bmatrix} x_d(t) \\ y_d(t) \\ \theta_d(t) \end{bmatrix} \tag{10}$$

Thus, one obtains

$$\begin{bmatrix} \dot{x}_e(t) \\ \dot{y}_e(t) \\ \dot{\theta}_e(t) \end{bmatrix}=\begin{bmatrix} \dot{x}(t) \\ \dot{y}(t) \\ \dot{\theta}(t) \end{bmatrix}-\begin{bmatrix} \dot{x}_d(t) \\ \dot{y}_d(t) \\ \dot{\theta}_d(t) \end{bmatrix}=P^{\#}(\theta(t))\begin{bmatrix} r\omega_1(t) \\ r\omega_2(t) \\ r\omega_3(t) \\ r\omega_4(t) \end{bmatrix}-\begin{bmatrix} \dot{x}_d(t) \\ \dot{y}_d(t) \\ \dot{\theta}_d(t) \end{bmatrix} \tag{11}$$

Similarly, the control goal is to find the motors' angular velocities $\begin{bmatrix} \omega_1(t) & \omega_2(t) & \omega_3(t) & \omega_4(t) \end{bmatrix}^T$ such that the closed-loop error system is globally asymptotical stable. In doing so, one proposes the following trajectory tracking law such that

$$\begin{bmatrix} \omega_1(t) \\ \omega_2(t) \\ \omega_3(t) \\ \omega_4(t) \end{bmatrix}=\frac{1}{r}P(\theta(t))\left(-K_p\begin{bmatrix} x_e(t) \\ y_e(t) \\ \theta_e(t) \end{bmatrix}-K_I\begin{bmatrix} \int_0^t x_e(\tau)d\tau \\ \int_0^t y_e(\tau)d\tau \\ \int_0^t \theta_e(\tau)d\tau \end{bmatrix}+\begin{bmatrix} \dot{x}_d(t) \\ \dot{y}_d(t) \\ \dot{\theta}_d(t) \end{bmatrix}\right) \tag{12}$$

where the matrices, K_p and K_I, are symmetric and positive definite. Substituting (12) into (11) leads to the underlying closed-loop error system governed by

$$\begin{bmatrix} \dot{x}_e(t) \\ \dot{y}_e(t) \\ \dot{\theta}_e(t) \end{bmatrix} = -K_p \begin{bmatrix} x_e(t) \\ y_e(t) \\ \theta_e(t) \end{bmatrix} - K_I \begin{bmatrix} \int_0^t x_e(\tau)d\tau \\ \int_0^t y_e(\tau)d\tau \\ \int_0^t \theta_e(\tau)d\tau \end{bmatrix} \qquad (13)$$

Similar to point stabilization, the Lyapunov function candidate can be chosen as equation (8), and from (9) one can easily prove that the closed-loop error system for trajectory tracking control is asymptotically stable. Worthy of mention is that the point stabilization and trajectory tracking control problems can be simultaneously achieved by the control law (12). The unified control law (12) becomes a point stabilization one if the desired pose $\begin{bmatrix} x_d(t) & y_d(t) & \theta_d(t) \end{bmatrix}^T$ can be either the time-dependent trajectory or the fixed destination posture.

3 Intelligent Motion Control Using ACO

3.1 Ant Colony Optimization

ACO algorithm is a general-purpose optimization technique based on a graph consisting of nodes and edges for solving various combinatorial problems. Optimization problems solutions can be expressed in terms of feasible paths on the graph. Among these feasible paths, the ACO algorithm aims to find the one with minimum performance index or cost.

In ACO algorithm, a colony of artificial ants is created to find solutions. At each generation, each ant completes a tour by choosing the nodes according to the transition rule. In order to prevent premature convergence, after completion of a path by each ant, pheromone intensities on links are evaporated with a pheromone update rule. With each edge, (i, j), of the graph is associated a total pheromone concentration, τ_{ij}. At each node, each ant executes a decision policy to determine the next link of the path. If ant k is currently located at node i, it selects the next node $j \in N_i^k$, based on the transition probability

$$\varphi^k{}_{ij}(t) = \begin{cases} \dfrac{\left[\tau_{ij}(t)\right]^\alpha \left[\eta_{ij}\right]^\beta}{\sum_{l \in N_i^k} \left[\tau_{il}(t)\right]^\alpha \left[\eta_{il}\right]^\beta}, & \text{if } j \in N_i^k \\ 0 & , \text{ if } j \notin N_i^k \end{cases} \qquad (14)$$

where N_i^k is the set of feasible nodes connected to node i, with respect to ant k. α and β are positive constant used to amplify the influence of pheromone concentrations. Note that the heuristic information η_{ij} is a problem-dependent function to be minimized given by

$$\eta_{ij} = \frac{1}{d_{ij}} \qquad (15)$$

where d_{ij} is the cost between the nodes i and j.

The pheromone update rule is given by

$$\tau_{ij}(t+1) = (1-\rho)\tau_{ij}(t) + \Delta\tau_{ij}(t) \tag{16}$$

with

$$\Delta\tau_{ij}(t) = \sum_{k=1}^{m} \Delta\tau^{k}_{ij}(t) \tag{17}$$

where $\Delta\tau^{k}_{ij}(t)$ is the amount of pheromone deposited by ant k on link (i, j) at time step t. m denotes the number of ants, $0 < \rho < 1$ is the pheromone decay parameter, namely that $(1-\rho)$ represents the evaporation rate. $\Delta\tau^{k}_{ij}(t)$ is given by

$$\Delta\tau^{k}_{ij}(t) = \begin{cases} \dfrac{Q}{L_k}, & \text{if } k^{th} \text{ ant uses edge}(i, j) \text{ in its tour} \\ 0, & \text{otherwise} \end{cases}$$

where Q is a positive constant and L_k is the tour cost of the kth ant. Global information is therefore used to update pheromone concentrations. This algorithm will terminate either when the maximum number of iterations is reached or an acceptable solution is found.

3.2   Application to Intelligent Motion Control

Although the kinematic controller for the four-wheeled omnidirectional mobile robot was synthesized in (12), the two control matrices K_P and K_I were not optimally chosen to obtain optimal performance. This subsection aims to employ the ACO computing method in Section 3.1 to design an optimal motion controller for omnidirectional mobile robots. The control parameters $K_P = diag\{k_{p1}, k_{p2}, k_{p3}\}$ and $K_I = diag\{k_{i1}, k_{i2}, k_{i3}\}$ in (12) are optimized via ACO algorithm to achieve trajectory tracking and stabilization for omnidirectional mobile robots.

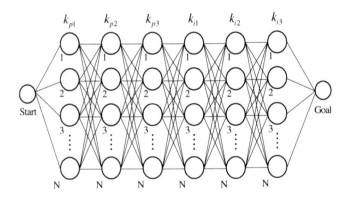

Fig. 2. ACO graph with controller parameters for omnidirectional mobile robots

The design of the ACO-based controller with K_p and K_I can be represented as a graph problem shown in Fig. 2 with controller parameter, $k_{p1}, k_{p2}, k_{p3}, k_{i1}, k_{i2}$ and k_{i3}. The searching range of these parameters is 0 to 100 with resolution 100/N. The performance index (fitness value) of the ACO-based controller is integral square error (ISE).

In ACO-based optimal controller design, each ant constructs a path and then deposits the pheromone. Using transition rule (14) and pheromone update rule (16), every ant completes their tours to find the best path. The ACO algorithm for finding the optimal motion controller parameters $k_{p1}, k_{p2}, k_{p3}, k_{i1}, k_{i2}$ and k_{i3} are summarized as follows.

Step 1. Initialization.
   (1) Set the iteration counter $t = 0$ and set the terminate iteration number.
   (2) Initialize the pheromone concentration on each link $\tau_{ij}(0)$ to a small random value.
Step 2. Define m (the number on ant), t_{max} (the maximum number of iterations), τ_0 (the initial pheromone concentration of each node) and ρ (the decay parameter).
Step 3. Each ant travels in the ACO-graph based on the probability in (14). If all ants complete their tour, calculate the cost (ISE) for each ant's tour.
Step 4. Update pheromone τ_{ij} using (16) for every edge (i, j).
Step 5. Check the stop criterion. If the stop criterion is not matched, go to Step 3 and set $t = t+1$, otherwise, output the optimal path and its corresponding controller parameter $k_{p1}, k_{p2}, k_{p3}, k_{i1}, k_{i2}, k_{i3}$ and stop the algorithm.

4 Simulation Results and Discussion

The aim of the simulations is to examine the effectiveness and performance of the proposed ACO-based kinematic control law (12) to the omnidirectional mobile platform. The number of ants in the ACO algorithm is 400. These simulations are performed with the following parameters: $L=23$cm, $r=5.08$cm, $\alpha = \beta = 1$, $\rho = 0.1$ and $Q=2$.

4.1 Point Stabilization

The first simulation was conducted to investigate the regulation performance of the proposed ACO control law (12). The initial pose of the omnidirectional mobile platform was assumed at the origin, i.e., $[x_0 \quad y_0 \quad \theta_0] = [0\text{ m} \quad 0\text{ m} \quad 0\text{rad}]$, and the desired final 8 goal postures are located on the unit circle, given by $\left[2\cos(\frac{n\pi}{4})\text{ m} \quad 2\sin(\frac{n\pi}{4})\text{ m} \quad \frac{\pi}{2}\text{ rad}\right]^T, n=0,1,...,7$. Fig. 3 depicts all the simulated trajectories of the omnidirectional mobile robot from the origin to the goal poses, and Fig. 4 shows the heading behavior of the proposed stabilization law for the platform moving towards the desired orientation $\pi/2$ in the case $n=1$. Through simulation results, the mobile robot with the proposed ACO stabilization method has been shown capable of reaching the desired postures. Fig. 5 presents the ISE of the ACO-based controller to achieve stabilization.

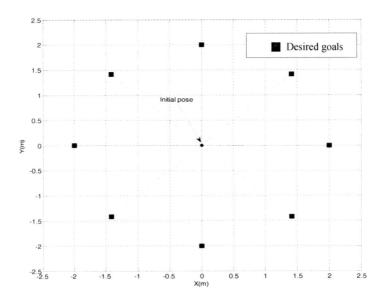

Fig. 3. Simulation results of the proposed ACO-based motion controller for achieving stabilization

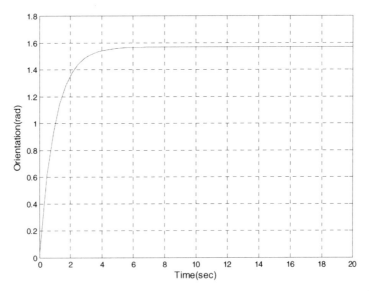

Fig. 4. Illustration of the orientation behavior moving towards the desired orientation of $\frac{\pi}{2}$ in the case $n=1$

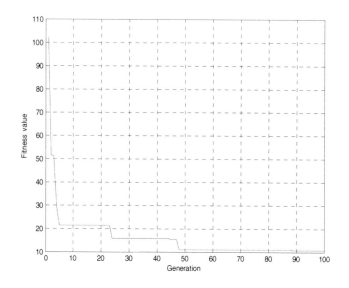

Fig. 5. Fitness value (ISE) of the proposed ACO-based controller to achieve stabilization

4.2 Elliptic Trajectory Tracking

The elliptic trajectory tracking simulation is aimed to explore how the proposed controller (12) steers the mobile platform to exactly track an elliptic trajectory described

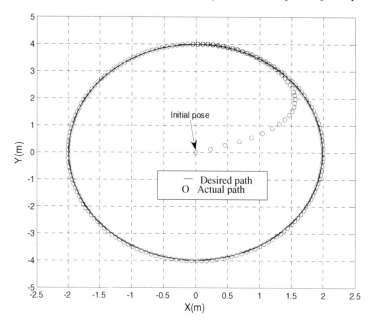

Fig. 6. Simulation result of the elliptic trajectory tracking

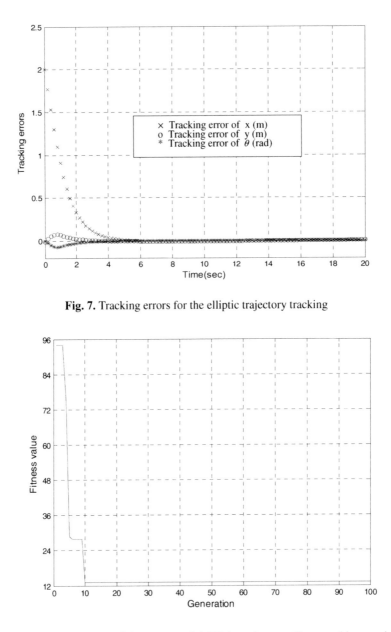

Fig. 7. Tracking errors for the elliptic trajectory tracking

Fig. 8. Performance index of the proposed ACO-based controller to achieve trajectory tracking

by $\left[x_r \quad y_r \quad \theta_r\right]=\left[2\cos w_r t(\text{m}) \quad 4\sin w_r t(\text{m}) \quad 0(\text{rad})\right]$, $w_r = 0.2 \text{ rad/sec}$. The simulation assumed that the platform got started at $\left[x_0 \quad y_0 \quad \theta_0\right]=\left[0\,\text{m} \quad 0\,\text{m} \quad 0\,\text{rad}\right]$. Fig. 6 presents the simulation result for elliptic trajectory tracking of the mobile robot. The tracking errors for the elliptical trajectory are depicted in Fig. 7. Fig. 8 presents the performance index of the proposed ACO controller to achieve trajectory tracking. These results indicate that the proposed ACO kinematic controller (12) is capable of successfully steering the omnidirectional mobile robot to track the elliptic trajectory.

5 Conclusion

This paper has presented an optimal motion controller based on ACO algorithm for four-wheeled omnidirectional mobile robots to achieve both trajectory tracking and stabilization. Based on the kinematic model, the optimial motion controller has been synthesized via ACO algorithm to achieve both trajectory tracking and stabilization. Through simulation results, the proposed ACO-based motion control method has been shown to achieve stabilization and trajectory tracking. An interesting topic for future work would be how to design an ACO-based dynamic controller for four-wheeled omnidirectional mobile robots.

Acknowledgment. The authors gratefully acknowledge financial support from the National Science Council, Taiwan, R.O.C., under grant NSC99-2628-E-241-003.

References

1. Kalmár-Nagy, T., D'Andrea, R., Ganguly, P.: Near-optimal dynamic trajectory generation and control of an omnidirectional vehicle. Robotics and Autonomous Systems 46(1), 47–64 (2004)
2. Chwa, D.K.: Sliding mode tracking control of nonholonomic wheeled mobile ro-bots in polar coordinates. IEEE Transactions on Control Systems Technology 12(4), 637–644 (2004)
3. Huang, H.C., Tsai, C.C.: FPGA implementation of an embedded robust adaptive controller for autonomous omnidirectional mobile platform. IEEE Transactions on Industrial Electronics 56(5), 1604–1616 (2009)
4. Jiang, Z.P., Nijmeijer, H.: A recursive technique for tracking control of non-holonomic systems in chained form. IEEE Transactions on Automatic Control 44(2), 265–279 (1999)
5. Lee, T.C., Song, K.T., Lee, C.H., Teng, C.C.: Tracking control of unicycle-modeled mobile robots using a saturation feedback controller. IEEE Transac-tions on Control Systems Technology 9(2), 305–318 (2001)
6. Li, T.H., Chang, S.J., Chen, Y.X.: Implementation of human-like driving skills by autonomous fuzzy behavior control on an FPGA-based car-like mobile robot. IEEE Transactions on Industrial Electronics 50(5), 867–880 (2003)
7. Pin, F.G., Killough, S.M.: A new family of omnidirectional and holonomic wheeled platforms for mobile robots. IEEE Transactions on Robotics and Automation 10(4), 480–489 (1994)
8. Byun, K.S., Kim, S.J., Song, J.B.: Design of a four-wheeled omnidirectional mo-bile robot with variable wheel arrangement mechanism. In: Proceeding of the 2002 IEEE International Conference on Robotics and Automation, pp. 720–725 (2002)
9. Purwin, O., D'Andrea, R.: Trajectory generation for four wheeled omnidirectional vehicles. In: Proceeding of 2005 American Control Conference, Portland, pp. 4979–4984 (2005)

10. Shing, C.C., Hsu, P.L., Yeh, S.S.: T-S fuzzy path controller design for the omnidi-rectional mobile robot. In: 32nd Annual Conference on IEEE Industrial Electronics, IECON 2006, Taiwan, pp. 4142–4147 (2006)
11. Li, T.H., Chen, C.Y., Hung, H.L., Yeh, Y.C.: A fully fuzzy trajectory tracking control design for surveillance and security Robots. In: Proceeding of 2008 IEEE In-ternational Conference on Systems, Man and Cybernetics (2008)
12. Tsai, C.C., Wang, T.Y., Wang, Z.C., Wu, Z.R.: Dynamic motion controller design for a four-wheeled omnidirectional mobile robot. In: Proceeding of the International Conference on Automation Technology (2009 Automation) (2009)
13. Manikas, T., Ashenayi, W.K., Wainwright, R.L.: Genetic algorithms for autono-mous robot navigation. IEEE Instrumentation & Measurement Magazine 10(6), 26–31 (2007)
14. Tang, K.S., Kim, F.M., Guanrong, C., Kwong, S.: An optimal fuzzy PID controller. IEEE Transactions on Industrial Electronics 48(4), 757–765 (2001)
15. Kan, J., Li, W., Liu, J.: Fuzzy immune self-tuning PID controller and its simulation. In: IEEE Conference on Industrial Electronics and Applications, pp. 625–628 (2008)
16. Jung, I.K., Hong, K.B., Hong, S.K., Hong, S.C.: Path planning of mobile robot using neural network. In: Proceedings of the IEEE International Symposium on Industrial Electronics, vol. 3, pp. 979–983 (1999)
17. Dorigo, M., Maniezzo, V., Colorni, A.: The ant system: optimization by a colony of cooperating agents. IEEE Transaction on Systems, Man, and Cybernetics-Part B 1, 1–13 (1996)
18. Sim, K.M., Sun, W.H.: Ant colony optimization for routing and load-balancing: survey and new directions. IEEE Transaction on Systems, Man, and Cybernetics-Part A 33(5), 560–572 (2003)
19. Watanabe, I., Matsui, S.: Improving the performance of ACO algorithms by adap-tive control of candidate set. In: Proceeding of the 2003 IEEE Evolutionary Computation, vol. 2, pp. 1355–1362 (2003)

Adaptive Sliding-Mode Speed Control for Electric Unicycle

Shui-Chun Lin

Department of Electronic Engineering, National Chin-Yi University of Technology,
Taichung, Taiwan, ROC
lsc@ncut.edu.tw

Abstract. This paper presents an adaptive hierarchical decoupling sliding-mode speed controller for an electric unicycle. A completely dynamic model of the electric unicycle moving in a flat terrain is derived using Lagrangian mechanics. With the model, an aggregated hierarchical sliding-mode control is used to accomplish robust self-balancing and velocity control (regulation) of the electric unicycle incorporating with viscous and static frictions. Computer simulations and experimental results are conducted for illustration of the effectiveness and applicability of the proposed control method.

Keywords: electric unicycle, Lagrangian mechanics, modeling, self-balancing, sliding-mode control.

1   Introduction

Electric unicycles have been considered by a kind of energy-saving vehicle for short-distance transportation for people shortly moving from one place to another. Recently, one-wheeled omnidirectional unicycle, called Honda U3-X, has been reported and demonstrated by Honda in [1], thus showing the possibility of being a kind of transportation vehicle, satisfying human demands of pollution-free, convenient and low-cost transportation. This one-wheeled vehicle is a reduced type of the two-wheeled transporter, thus simplifying its hardware structure. The vehicle can be also regarded as a special wheeled mobile inverted pendulum that can be balanced by powering wheel(s) to achieve self stabilization. Forward and backward movements are achieved by the rider's intensions to change the position of his/her center of gravity (COG), and the yaw motion is accomplished by body movement.

From the viewpoint of control technology, the electric unicycle is an inherently unstable and highly nonlinear system, requiring a highly complicated modeling and control policy. Recently, unicycles have attracted considerable attention in both academia and industry [2]-[7]. Some researchers considered that electric unicycles can be regarded as a kind of vehicle for future transportation [8]. Such unicycles are indeed a pollution-free and extremely compact transportation vehicle driven by only one DC motor. The kind of transporter can be easily constructed by a synthesis of mechatronics, control and software. For example, the researchers in [8] presented a kind of low-tech electric unicycle which was made by the off-the-shelf inexpensive components, such as

T.-H.S. Li et al. (Eds.): FIRA 2011, CCIS 212, pp. 107–115, 2011.
© Springer-Verlag Berlin Heidelberg 2011

one DC brush motor with a cheap gearbox, a rubber-based wheel, a eight-bit micro-controller, a PWM-based motor driver, a rate gyro, an accelerator, and nickel metal hydride (NiMH) batteries.

Motivated by the Segway TM, a two-wheeled self-balancing vehicle, some re-searchers in [8] found that the unicycle can be simplified from the Segway by em-ploying only one driving wheel, thus considering that the unicycle can be modeled as a one-wheeled mobile inverted pendulum. This idea can be successful if the rider can skillfully balance the unicycle without lateral falling. Once the idea works, several stabilization concepts proposed by Oryschuk *et. al.* [9], Grasser *et al.* [10], Pathak *et al.* [11], Zhao *et al.* [12], Conceicao *et al.* [13], Low *et al.* [14], Damien, *et al.* [15], Liaw *et al.* [16], could then be applied to control the unicycle. Following these studies, the authors in [17] constructed an experimental two-wheeled self-balancing vehicle with low-cost and low-tech components, and proposed one adaptive controller for the ve-hicle with parameter variations and both coulomb and static frictions, in order to achieve self-balancing and speed control. Indeed, the control of the electric unicycle can be thought of as an under-actuated control problem, which has been investigated by several researchers [18, 19, 20]. In particular, Lo and Kuo [18] provided a decoupled sliding-mode control to stabilize a nonlinear system with four state variables, Lin and Mon [19] offered a hierarchical decoupling sliding-mode control to regulate a more general class of under-actuated control systems, and Wang *et al.* [20] presented two systematic sliding-mode design methods, called incremental hierarchical structure sliding-mode control and aggregated hierarchical structure sliding-mode control, for a class of under-actuated mechanical systems. However, these approaches [18, 19, 20] have not been applied to the electric unicycle yet!

The objectives of this paper are to develop methodologies for dynamic modeling and aggregated hierarchical sliding-mode control of the electric unicycle. Two contribu-tions of the paper are delineated as follows. First, a new dynamic model of the electric unicycle with viscous and static frictions is derived using Lagrangian mechanics. Second, an aggregated hierarchical sliding-mode controller is developed to achieve robust self-balancing and speed tracking of the electric unicycle.

The rest of the paper is organized as follows. Section 2 is devoted to establishing the dynamic model of the electric unicycle with one brushless motor. In Section 3, the sliding-mode controller is synthesized to achieve the design goals. Several simulations and experimental results are respectively performed in Section 4 and 5 to illustrate the effectiveness of the proposed control method. Section 6 concludes the paper.

2   Uncertain System Model

The section is aimed at deriving the mathematical model of an electric unicycle. As Fig.1 shows, the working principle of the unicycle is interpreted as follows. If the rider leans forward, the unicycle will move forward in order to maintain the rider's body without falling. Similarly, if the rider leans backward, then the unicycle will move backwards for balancing. After understanding the basic working principle of the vehi-cle, one desires to derive the dynamic model of the rider mounting the unicycle using Lagrangian mechanics. Note that the modeling process is based on a by two major components: the body and the wheel, as Fig.2 shows. All parameters required for model

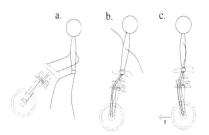

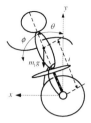

Fig. 1. Illustration of riding an electric unicycle

Fig. 2. Free-body diagrams of the body (including the rider) and the wheel

derivation of the unicycle are: ϕ the rotation angle, θ the tilt angle, $\bar{\theta}_x$ and $\bar{\theta}_y$ the reference frame in both and y axes, $\bar{\theta}_i$ the rotation vector expressed by (4), I_m, m and $\dot{x}$ the moment of inertia, mass, and velocity of the wheel, I_M and M the moments of inertia and mass of the body except the wheel, $\bar{i}, \bar{j}$ and $\bar{k}$ the three normal unit vectors in Cartesian coordinates, l the distance between COG of the body and the center of the wheel, M_p the mass of the chassis with the weight of the rider, r the radius of the wheel, $\bar{v}_G$ the velocity of the body, v_s the state vector of the electric coefficients, c_θ and c_ϕ the static friction coefficients, and τ the torque applied on the unicycle, T_{wheel} the kinetic energy of the wheel, T_{body} the kinetic energy the body, V_{wheel} the potential energy of the wheel, V_{body} the potential energy of the body, μ_θ and μ_ϕ the viscous friction wheel. To describe the simplification assumption that the rider on the unicycle is constructed dynamics of the unicycle, one employs the reference frame $\bar{\theta}_x$, $\bar{\theta}_y$ together with the tilt angle θ to orient the body [2], and the rotation angle ϕ of the wheel, as shown in Fig.2. Before proceeding with model derivation, it is necessary to obtain the kinetic and potential energies of the wheel and the body. For the wheel with the hub motor one obtains

$$T_{wheel} = \left(m_2 \dot{x}^2\right)/2 + \left(I_m (\dot{\phi}+\dot{\theta})^2\right)/2$$
$$V_{wheel} = 0 \tag{1}$$

where I_m, m and $\dot{x}$ are, respectively the moment of inertia, mass, and velocity of the wheel. Assuming that no slip occurs between the wheel and the floor, the kinetic energy of the wheel can be rewritten as

$$T_{wheel} = \left(m_2 r^2 (\dot{\phi}+\dot{\theta})^2 + I_m (\dot{\phi}+\dot{\theta})^2\right)/2 = \left((m_2 r + I_m)/2\right)(\dot{\phi}+\dot{\theta})^2 \tag{2}$$

The velocity of the center of gravity (COG) for the body can be expressed by

$$v_x = r(\dot{\phi}+\dot{\theta}) + l\dot{\theta}\cos(\theta) \tag{3}$$

where l is the distance between COG of the body and the center of the wheel. Similarly, the velocity expression for COG of the body in the y axis is given as follows;

$$v_y = -l\dot{\theta}\sin(\theta) \tag{4}$$

For the body it follows that

$$T_{body} = M(v_x^2 + v_y^2)/2 + I_M\dot{\theta}^2/2 \tag{5}$$

$$V_{body} = Mgl\cos(\theta) \tag{6}$$

where I_M and M are respectively the moments of inertia and mass of the body. Note that the mass M includes the mass of the rider and the unicycle except the wheel with the hub motor. Therefore, the total kinetic energy of the electric unicycle is

$$T_T = I_\phi(\dot{\phi}+\dot{\theta})^2/2 + I_\theta\dot{\theta}^2/2 + \beta\dot{\theta}(\dot{\theta}+\dot{\phi})\cos(\theta) \tag{7}$$

Where

$$I_\phi = m_2 r^2 + I_m + Mr^2, \quad \beta = m_1 rl \tag{8}$$

and

$$I_\theta = I_M + Ml^2 \tag{9}$$

Define the state vector of the electric unicycle by

$$v_s = \begin{bmatrix} \theta & \phi \end{bmatrix}^T \tag{10}$$

and model the static and viscous frictions by using the subsequent vector

$$D(\dot{v}_s) = \begin{bmatrix} 0 & \mu_\phi\dot{\phi} + c_\phi\,\mathrm{sgn}(\dot{\phi}) \end{bmatrix}^T \tag{11}$$

where $\mathrm{sgn}(\cdot)$ means the signum function, and the frictions between body-wheel and wheel-ground are modeled by four coefficients; μ_ϕ is the viscous coefficients; c_ϕ is the coefficients of the static frictions. Let τ be the torque applied to the wheel in the direction $\bar{k}$.

Next, the mathematical model of the vehicle can be derived by defining its Lagrangian function L as $L = T_T - V_{body}$, and then by using the Euler-Lagrange equations. Hence, the equations of motion for electric unicycle are obtained from

$$\frac{d}{dt}\left(\frac{\partial L}{\partial \dot{v}_s}\right) - \frac{\partial L}{\partial v_s} = \begin{bmatrix} 0 \\ \tau \end{bmatrix} - D(\dot{v}_s) \tag{12}$$

which, after manipulating the derivatives in the Euler-Lagrange equations, leads to

$$(I_\phi + I_\theta + 2\beta\cos(\theta))\ddot{\theta} + (I_\phi + \beta\cos(\theta))\ddot{\phi} - \beta\dot{\theta}^2\sin(\theta) - \frac{\beta g}{r}\sin(\theta) = 0 \tag{13}$$

$$(I_\phi + \beta\cos(\theta))\ddot{\phi} + I_\phi\ddot{\theta} - \beta\dot{\theta}^2\sin(\theta) = \tau - \mu_\phi\dot{\phi} - c_\phi\mathrm{sgn}(\dot{\phi}) \tag{14}$$

Combining (13) and (14) gives two second-order dynamic equations of the inclination and rotation angles controlled by the torque τ.

$$\ddot{\theta} = A(\theta)(\tau - \mu_\phi \dot{\phi} - c_\phi \mathrm{sgn}(\dot{\phi})) + B(\theta, \dot{\theta}) \tag{15}$$

$$\ddot{\phi} = C(\theta)(\tau - \mu_\phi \dot{\phi} - c_\phi \mathrm{sgn}(\dot{\phi})) + D(\theta, \dot{\theta}) \tag{16}$$

where

$$A(\theta) = -\left(I_\phi + \beta \cos(\theta)\right)/\delta \tag{17}$$

$$B(\theta, \dot{\theta}) = \left\{m_1 l g I_\phi \sin(\theta) - \beta^2 \dot{\theta}^2 \cos(\theta)\sin(\theta)\right\}/\delta \tag{18}$$

$$C(\theta) = \left(I_\theta + I_\phi + 2\beta \cos(\theta)\right)/\delta \tag{19}$$

$$D(\theta, \dot{\theta}) = \left[\left((I_\theta + \beta \cos(\theta))\beta\dot{\theta}^2 \sin(\theta)\right) - \left((I_\phi + \beta \cos(\theta))m_1 l g \sin(\theta)\right)\right]/\delta \tag{20}$$

where $\delta = I_\theta I_\phi - [\beta \cos\theta]^2$. Note that both variables $\ddot{\theta}$ and $\ddot{\phi}$ are simultaneously controlled by the same torque τ; this reveals that the electric unicycle is indeed an under-actuated control system. Moreover, the values of I_θ, I_ϕ, M depends heavily upon the weight of the rider, namely that they will vary with the rider's weight. Thus, I_θ, I_ϕ, M can be decomposed into their nominal values, $I_{\theta 0}, I_{\phi 0}, M_0$, and their perturbed terms, ΔI_θ, ΔI_ϕ and ΔM, such the $I_\theta = I_{\theta 0} + \Delta I_\theta, I_\phi = I_{\phi 0} + \Delta I_\phi$, and $M = M_0 + \Delta M$. Thus,

$$A(\theta) = A_0(\theta) + \Delta A(\theta) \tag{21a}$$

$$B(\theta, \dot{\theta}) = B_0(\theta, \dot{\theta}) + \Delta B(\theta, \dot{\theta}) \tag{21b}$$

$$C(\theta) = C_0(\theta) + \Delta C(\theta) \tag{21c}$$

$$D(\theta, \dot{\theta}) = D_0(\theta, \dot{\theta}) + \Delta D(\theta, \dot{\theta}) \tag{21d}$$

where $A_0(\theta) = -\left(I_{\phi 0} + \beta_0 \cos(\theta)\right)/\delta_0$, and $\delta_0 = I_{\theta 0} I_{\phi 0} - [\beta_0 \cos(\theta)]^2$, $\beta_0 = M_0 l_0 r_0$,
$\left|\Delta A(\theta) = A(\theta) - A_0(\theta)\right| \le K_A < \infty, B_0(\theta, \dot{\theta}) = \left\{\beta_0 g I_{\phi 0} \sin(\theta)/r - \beta_0^2 \dot{\theta}^2 \sin(\theta)\cos(\theta)\right\}/\delta_0$
and $\left|\Delta B(\theta, \dot{\theta}, \dot{\phi}) = B(\theta, \dot{\theta}, \dot{\phi}) - B_0(\theta, \dot{\theta}, \dot{\phi})\right| \le K_B < \infty$,
$C_0(\theta) = \left(I_{\theta 0} + I_{\phi 0} + 2\beta_0 \cos(\theta)\right)/\delta_0$, $\left|\Delta C(\theta) = C(\theta) - C_0(\theta)\right| \le K_C < \infty$
$D_0(\theta, \dot{\theta}) = \left\{(I_{\theta 0} + \beta_0 \cos(\theta))\beta_0 \dot{\theta}^2 \sin(\theta)\right\}/\delta_0 - \left\{(I_{\phi 0} + \beta_0 \cos(\theta))\beta_0 g \sin(\theta)/r\right\}/\delta_0$ and
$\left|\Delta D(\theta, \dot{\theta}) = D(\theta, \dot{\theta}, \dot{\phi}) - D_0(\theta, \dot{\theta}, \dot{\phi})\right| \le K_D < \infty$. With (21), (15) and (16) turn out

$$\ddot{\theta} = A_0(\theta)\tau + B_0(\theta, \dot{\theta}) + \Delta_\theta, \Delta_\theta = \Delta A(\theta)\tau + \Delta B(\theta, \dot{\theta}) \tag{22}$$

$$\ddot{\phi} = C_0(\theta)\tau + D_0(\theta, \dot{\theta}) + \Delta_\phi, \Delta_\phi = \Delta C(\theta)\tau + \Delta D(\theta, \dot{\theta}) \tag{23}$$

where both perturbed terms Δ_θ and Δ_ϕ are assumed to be bounded under the assumption of boundedness of the applied torque τ.

3 Adaptive Controller Synthesis

This section is devoted to designing an adaptive decoupling sliding-mode controller for achieving constant velocity control of the electric unicycle. This control goal can be reduced to the underactuated regulation problem, namely that θ is stabilized at zero, i.e., $\theta_{desired} = 0$, and the desired constant angular velocity is maintained at $\dot{\phi}_{desired}$ using only one actuator. Note that if the controller works well, then the unicycle move at the speed of $r\dot{\phi}_{desired}$ where r is the radius of the wheel. In what follows, an adaptive decoupling sliding-mode control method is proposed to achieve such a design goal. The design procedure is divided into two steps; the first step designs an adaptive decoupling sliding mode controller, and the second step addresses its stability issue.

3.1 Adaptive Aggregated Hierarchical Sliding-Mode Speed Control

To stabilize the electric unicycle in self-balancing and velocity control, one constructs two first-layer sliding surface functions proposed in [25].

$$S_{TH}(\theta) = \dot{\theta} + k_{TH}(\theta - \theta_{desired}) = \dot{\theta} + k_{TH}\theta \tag{24}$$

$$S_{PH}(\dot{\phi}) = \dot{\phi} - \dot{\phi}_{desired} \tag{25}$$

where k_{TH} is the positive constant; $\dot{\phi}_{desired}$ is the constant goal velocity, i.e., $\dot{\phi}_{desired} = v_{desired}/r$. Note that the desired inclination $\theta_{desired}$ is always maintained at zero. Next, define the second-layer sliding surface as

$$S_T = S_{TH}(\theta) + \alpha S_{PH}(\dot{\phi}) \tag{26}$$

where the constant parameter α is a real constant. Differentiating S_T and using (15) and (16) give

$$\dot{S}_T = \left(A_0(\theta) + \alpha C_0(\theta)\right)\left(\tau - \mu_\phi \dot{\phi} - c_\phi \mathrm{sgn}(\dot{\phi})\right) + \left(B_0(\theta,\dot{\theta},\dot{\phi}) + k_{TH}\dot{\theta} + \alpha D_0(\theta,\dot{\theta},\dot{\phi})\right) + \Delta F \tag{27}$$

where $\Delta F = \Delta_\theta + \alpha \Delta_\phi$ is bounded. Moreover, it is reasonably assumed that $|\Delta F| \le K_F$ where K_F is real, constant and positive. From (27), it is reasonable to propose the following adaptive control law.

$$\tau = -\frac{B_0(\theta,\dot{\theta},\dot{\phi}) + k_{TH}\dot{\theta} + \alpha D_0(\theta,\dot{\theta},\dot{\phi})}{A_0(\theta) + \alpha C_0(\theta)} - \frac{\beta S_T + \hat{k}_v \,\mathrm{sgn}(S_T)}{A_0(\theta) + \alpha C_0(\theta)} + \hat{\mu}_\phi \dot{\phi} + \hat{c}_\phi \,\mathrm{sgn}(\dot{\phi}) \tag{28}$$

where $\mathrm{sgn}(\cdot)$ means the signum function; β is a positive real constant; $\hat{k}_v$ is the estimate of the constant parameter K_F; $\hat{\mu}_\phi$ and $\hat{c}_\phi$ are two estimates of both constant friction coefficients, μ_ϕ and c_ϕ. Moreover, the parameter updating laws for $\hat{k}_v$, $\hat{\mu}_\phi$ and $\hat{c}_\phi$ are given by

$$\dot{\hat{k}}_v = \gamma_1 |S_T| \tag{29a}$$

$$\dot{\hat{\mu}}_\phi = -\gamma_2 S_T \left(A_0(\theta) + \alpha C_0(\theta) \right) \dot{\phi} \tag{29b}$$

$$\dot{\hat{c}}_\phi = -\gamma_3 S_T \left(A_0(\theta) + \alpha C_0(\theta) \right) \operatorname{sgn}(\dot{\phi}) \tag{29c}$$

where γ_1, γ_2 and γ_3 are three positive adaptation gains.

3.2 Stability Analysis

In order to show that the second-layer sliding surface S_T converges to zero, the following Lyapunov function is proposed as

$$V = \left(S_T^2 / 2 \right) + \left(\tilde{k}_v^2 / 2\gamma_1 \right) + \left(\tilde{\mu}_\phi^2 / 2\gamma_2 \right) + \left(\tilde{c}_\phi^2 / 2\gamma_3 \right) \tag{30}$$

where $\tilde{k}_v = K_F - \hat{k}_v$, $\tilde{\mu}_\phi = \mu_\phi - \hat{\mu}_\phi$ $\tilde{c}_\phi = c_\phi - \hat{c}_\phi$. From (28)-(30), taking the time derivative of V yields

$$\dot{V}(t) = S_T \dot{S}_T - \left(\tilde{k}_v \dot{\hat{k}}_v / \gamma_1 \right) - \left(\tilde{\mu}_\phi \dot{\hat{\mu}}_\phi / \gamma_2 \right) - \left(\tilde{c}_\phi \dot{\hat{c}}_\phi / \gamma_1 \right) = -\beta S_T^2 \leq 0 \tag{31}$$

Since both $\dot{V}(t)$ is negative semidefinite, it is easy to prove via Lyapunov stability theory that the second-layer sliding function, S_T, converges to zero as time approaches infinity. Furthermore, the two first-layer sliding functions, S_{TH} and S_{PH}, can be shown to converge to zeros asymptotically using the aggregated hierarchical sliding mode control method in [25].

These indicate that the proposed control method can control the electric unicycle to reach its desired velocity and maintain θ at zero. The main result is summarized as below.

Theorem 1. Consider the electric unicycle's uncertain dynamic model (22)-(23) with the proposed adaptive hierarchical decoupling sliding-mode control law (29) with the parameter adjustment update laws (30a)-(30c). Then the second-layer sliding function converges to zero, namely that $S_T \to 0$, and the first-layer functions also tend to zero as time goes to infinity, .i.e., S_{TH} and $S_{PH} \to 0$ as $t \to \infty$. Moreover, $\theta \to 0$ and $\dot{\phi} \to \dot{\phi}_{desired}$ as $t \to \infty$.

4 Simultaions and Discussion

This section is devoted to conducting a simulation to examine the performance and merit of the proposed controller (28) with the parameter adjustment rules (30a-c). The first simulation adopts the subsequent parameters: the weights are $M = 78 \, \text{kg}$ and $m = 2 \, \text{kg}$, the moments of inertia are $I_\theta = 155.95 \, \text{Nm}$ and $I_\phi = 25.126 \, \text{Nm}$, and the radius of the wheel is $r = 0.2 \, \text{m}$, the static and viscous coefficients are respectively set by $\mu_\phi = 5$ and $c_\phi = 1$. The initial conditions are $\theta = 0.1745\,\text{rad} = 10\,\text{deg}$ and $\dot{\phi} = 0\,\text{rad/sec}$. The parameters of the proposed controller are set by $K_{TH} = 20$, $\alpha_0 = 0.1$, $\beta = 1$, $\gamma_1 = 0.1$ and $\gamma_2 = \gamma_3 = 0.01$, and the desired vehicle speed is 1 m/sec, i.e., $\dot{\phi}_{desired} = 5$ rad/sec. The simulation is done by Matlab/Simulink to steer the electric unicycle for self-balancing and constant speed tracking, namely that the tilt angle and the vehicle speed are respectively maintained at zero and 1 m/sec. Figs.3 depicts that the three sliding functions, S_T, S_{TH} and S_{PH}, approach zero, θ is maintained at zero in 2.5 seconds, and the velocity tended to 1.0 m/sec.

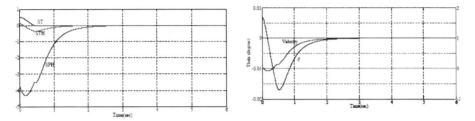

Fig. 3. (a). Time histories of the sliding surfaces S_T, S_{TH} and S_{PH} (b) Simulation results of velocity and θ of the proposed controller with self-balancing, with mass of cylinder is 80Kgw and radius of wheel is 0.2

5   Conclusions

This paper has presented an adaptive hierarchical decoupling sliding-mode controller for an electric unicycle accomplished by activating one brushless motor. The completely dynamic model of the electric unicycle moving in a flat terrain has been derived based on Lagrangian mechanics. The adaptive hierarchical decoupling sliding–mode controller has been proposed to accomplish robust balancing and velocity control of the electric unicycle by neglecting all of the friction coefficients. The effectiveness of the proposed modeling and control method has been exemplified by conducting numerical simulations and experimental results on the electric unicycle. An interesting topic for future research would be to implement the proposed control scheme using interval type II fuzzy neural networks and then conduct their experimental results.

Acknowledgments. The authors deeply acknowledge final support in part from National Science Council, Taiwan, ROC, under contract NSC98-2218-E-167-003.

References

1. Honda Com. (2009), http://www.honda.co.jp/news/2009/c090924.html
2. Sheng, Z., Yamafuji, K.: Postural stability of a human riding a unicycle and its emulation by a Robot. IEEE Trans. Robot. and Auto. 13(5), 709–720 (1997)
3. Brown Jr, H.B., Xu, Y.: A single-wheel, gyroscopically stabilized robot. IEEE Robotics & Automation Magazine 4(3), 39–44 (1997)

4. Pathak, K., Agrawal, S.K.: An integrated path-planning and control approach for non-holonomic unicycles using switched local potentials. IEEE Trans. Robotics and Automation 21(6), 1201–1208 (2005)

5. Cedervall, S., Hu, X.: Nonlinear Observers for Unicycle Robots with Range Sensors. IEEE Trans. Automatic Control 52(7), 1325–1329 (2007)

6. Jin, Z., Zhang, G.: the nonholonomic motion planning and control of the unicycle mobile robot. Proc. Intelligent Control and Automation (1), 3461–3465 (2006)

7. Colli, V.D., Tommasi, G., Scarano, M.: Single wheel longitudinal traction control for electric vehicles. IEEE Trans. Power Electronics 21(3), 799–808 (2006)

8. Ricky, G.: The Electric Unicycle Riding (aunicycle is the most useless thing you can do,) (2007), http://tlb.org/eunicycle.html

9. Oryschuk, P., Salerno, A., Al-Husseini, A.M., Angeles, J.: Experimental Validation of an Underactuated Two-Wheeled Mobile Robot. IEEE/ASME Trans. on Mechatronics 14(2), 252–257 (2009)

10. Grasser, F., Arrigo, A.D., Colombi, S.: JOE: A Mobile, Inverted Pendulum. IEEE Trans. Indus. Elec. 49(1), 107–114 (2002)

11. Pathak, K., Franch, J., Agrawal, S.K.: Velocity and position control of a wheeled inverted pendulum by partial feedback linearization. IEEE Trans. Robotics and Automation 21(3), 505–513 (2005)

12. Zhao, D., Deng, X., Yi, J.: Motion and Internal Force Control for Omnidirectional Wheeled Mobile Robots. IEEE/ASME Trans. on Mech. 14(3), 382–387 (2009)

13. Conceicao, A.S., Moreira, A.P., Costa, P.J.: Practical Approach of Modeling and Parameters Estimation for Omnidirectional Mobile Robots. IEEE/ASME Trans. on Mechatronics 14(3), 377–381 (2009)

14. Low, C.B., Wang, D.: GPS-Based Tracking Control for a Car-Like Wheeled Mobile Robot With Skidding and Slipping. IEEE/ASME Trans. on Mechatronics 13(4), 480–484 (2008)

15. Damien, L.D., Grand, C., Faiz, B.A., Guinot, J.C.: Doppler-Based Ground Speed Sensor Fusion and Slip Control for a Wheeled Rover. IEEE/ASME Trans. on Mechatronics 13(4), 484–492 (2009)

16. Liaw, H.C., Shirinzadeh, B.: Neural Network Motion Tracking Control of Piezo-Actuated Flexure-Based Mechanisms for Micro-/Nanomanipulation. IEEE/ASME Trans. on Mechatronics 14(5), 517–527 (2009)

17. Tsai, C.C., Huang, H.C., Lin, S.C.: Adaptive Neural Network Control of a Self-balancing Two-wheeled Scooter. IEEE Trans. on Industrial Electronics 57(4), 1420–1428 (2010)

18. Lo, J.C., Kuo, Y.H.: Decoupled fuzzy sliding-mode control. IEEE Trans. on Fuzzy Systems 6(3), 426–435 (1998)

19. Lin, C.M., Mon, Y.J.: Decoupling Control by hierarchical fuzzy sliding-mode controller. IEEE Trans. on Control System Technology 13(4), 593–598 (2005)

20. Wang, W., Liu, X.D., Yi, J.Q.: Structure design of two types of sliding-mode controllers for a class of under-actuated mechanical systems. IET Proc. of Control Theory and Applications 1(1), 163–172 (2007)

Dynamic Patrol Planning in a Cooperative Multi-robot System

Kao-Shing Hwang[1], Jin-Ling Lin[2], and Hui-Ling Huang[3]

[1,3] Department of Electrical Engineering, National Chung Cheng University, Chiayi, Taiwan
hwang@ccu.edu.tw
[2] Department of Information Management, Shih Hsin University, Taipei, Taiwan
jllin@cc.shu.edu.tw

Abstract. A cooperative multi-robot system is proposed to solve the problem of dynamic patrol planning. Each mobile robot has its own patrol mission in the beginning. The patrol mission of each robot needs to be updated when the number of mobile robots increases or decreases during patrol. From the results of the simulation, it is clear that the proposed approach demonstrates several advantages, such as decreased time complexity, a lower routing path cost, improved balance of workload among robots, the potential to scale to a large number of robots, and adaptability with regards to environmental perturbations introduced by changes in the number of robots in patrol.

Keywords: dynamic patrol planning, cooperative patrol planning, multi-robot system.

1 Introduction

Cooperative multi-robot/multi-agent systems have received increased attention due to their empirically demonstrated performance and advantages, such as robustness with regards to environmental perturbations or individual robot failure, as well as their scalability to a large number of robots [1][2][3][4][5][6][7][8]. However, little work has been done to investigate ways of giving such systems the capability of achieving a desired division of labor over a set of dynamically evolving concurrent tasks. This capability could help increase the efficiency and robustness of overall task performance, as well as open new domains where these systems could be regarded as a viable alternative to more complex control solutions. In this paper, a method for achieving the desired division of labor, such as robot patrol, is presented. The problem of multi-robot patrol has been investigated over the past few years. This subject is of interest for a number of reasons, the main one being its applicability to various domains, for example, post checking or goods delivery. For such problems, a team of robots is required to visit the posts, which should be visited exactly once, while monitoring the activity in order to detect changes in the state of the environment. Such systems of multiple robots engaged together in order to patrol a varying environment have been studied in various contexts. The problem of patrol planning [2][6], where a team of robots is required to visit a target point once, is

T.-H.S. Li et al. (Eds.): FIRA 2011, CCIS 212, pp. 116–123, 2011.

related to vehicle routing, traveling salesman [1], and so on, which all are the problems that are NP-Hard [1][2][3][6].

In this work, two patrol situations, with increasing and decreasing numbers of robot, are discussed. Robots in situations where spontaneous situations occur, such as the failure or reinforcement of robot force, are considered. Then a robust algorithm is derived to find the optimal patrol sequence for each on-duty robot in different dynamic environments. The proposed dynamic patrol planning based on the cooperative auction system has the advantages of shorter accumulated patrol distance, less computation complexity, and adaptability for various numbers of robots.

The paper is organized as follows. The following section presents the research background. Dynamic patrol planning is described in Section 3. Section 4 presents the simulations and discussion. Conclusions are provided in the final section.

2 Research Background

MINISUM is used to evaluate the performance of the algorithm derived in this study [3][6]. Here MINISUM means that the total energy that robots consume should be reduced as much as possible. In other words, the goal of cooperation for a multi-robot system is to minimize the total length of paths that all of the robots patrol. Let T_i denote the set of patrol points for robot i, $Robot_i$ to patrol and $T = \{T_1, T_2, \ldots, T_{|group\_mobile\_robot|}\}$, a partition of the set of patrol points, where $|group\_mobile\_robot|$ denotes the number of robots. Then the mathematic model of MINISUM can be defined as follows:

$$MINISUM : \min_{T} \sum_{i=1}^{|group\_mobile\_robot|} Patrol\_Path\_Length(Robot_i, T_i) \cdot \quad (1)$$

Where function $Patrol\_Path\_Length(Robot_i, T_i)$ will return the total distance of patrol after $Robot_i$ visits all of the patrol points in T_i.

When the number of robots changes, either decreases or increases, during the patrol, the robots' task needs to be adjusted in order to complete the patrol or enhance the performance of the group. There are two ways to deal with a rush job: revised assignment or direct assignment. Revised assignment will redo the assignment procedure and direct assignment will rearrange the unvisited patrol points only. The former is easier but more time-consuming, whereas the latter is more efficient, although its algorithm is more complicated. The proposed algorithm is the latter and performs during robots' patrol.

For the proposed system to begin work, it needs the following information.

- The number of points that need to be patrolled, denoted by $|P|$, where P denotes the set of patrol points.
- The location of each patrol point, denoted by $Loc(P_i)$, where P_i is the patrol point i.
- The number of robots that can currently patrol, denoted by $|R|$, where R denotes the set of mobile robots.
- The current location of each robot, denoted by $Loc(R_i)$, where R_i is the robot i
- The patrol path for each robot.

In addition to the input information, the proposed system assumes

- There is no barrier between robots and patrol points. In other words, Euclidean distance is used to represent the length of paths between static patrol points and moving robots.
- There is no collision problem while robots are patrolling.
- The robots are identical and do not have capacity constraints.
- The robots know their own location and the target locations.

3   Dynamic Patrol Planning

Sequential Single-Item Auction (SSIA) [9], Cooperative Auction Directed by Winner (CAD_W), and Cooperative Auction Directed by Loser (CAD_L) [10] are used to generate the patrol path for each robot in static environment. Since none of them guarantees that the task will be efficiently performed during the patrol since the robots may be out of order during the patrol even though we know the patrol points for each robot. On the other hand, the malfunctioning robots can join the task again if they have been repaired. Therefore, a dynamic patrol planning (DPP), based on the CAD_W (DPP_CAD_W) and CAD_L (DPP_CAD_L), is proposed to solve this kind of problem caused by an increasing or decreasing number of robots during the patrol.

Since direct assignment considers the unvisited patrol points only, it is more efficient than revised assignment and it is a better candidate for dynamic patrol planning. The direct assignment approach is based on the current patrol situation of robots. In other words, only the unvisited patrol points will be considered for patrol. SSIA needs to evaluate all of the patrol points in order to offer appropriate prices during auctions. Direct assignment requires robots to offer prices for these unvisited patrol points only. Therefore, SSIA is not suitable for providing solutions to dynamic patrol planning when the number of robots decreases or increases. In the CAD_W and CAD_L system, however, during each auction robots can offer appropriate prices even though they can only consider patrol points that are assigned to them and still unvisited. Dynamic patrol planning, therefore, can be easily performed by the CAD_W and CAD_L with slight modifications. The set of working robots, $R_{dynamic}$, includes the newly entering robots and excludes the malfunctioning robots. The current locations of robots become the starting locations. The CAD_W and CAD_L are then utilized, as discussed in the previous work [10], based on $R_{dynamic}$.

3.1   DPP When Some Working Robots Drop Out of the Patrol Task

With direct assignment, when the number of robots decreases, patrol points will still not have been assigned to robots. When any robots, R_{out}, cannot proceed with their task during the patrol, the patrol points, P_{out}, which had been assigned to R_{out} but have not yet been visited, are candidates for re-auction. In other words, P_{out} are the patrol points that need to be auctioned. Any robot in $R_{dynamic} = R - R_{out}$, which are still working, contains only unvisited patrol points now, $T_i = T_i - \{j \mid j \in T_i$ and had been visited$\}$, where T_i denotes the patrol points that had been assigned to robot i, though it now contains unvisited ones only. The current locations of robots become the starting locations and the auction continues until all patrol points in P_{out} have been re-assigned to robots in $R_{dynamic}$. In other words, dynamic patrol planning when some working robots have dropped out of the task is simply auctioning the patrol points P_{out} that have not yet been assigned to any robot.

3.2   DPP When Other Robots Join the Patrol Task

When robots, R_{in}, join the task during patrol the workload of the functioning robots, R, can be reduced because the unvisited patrol points in R can be auctioned again and assigned to these new robots in R_{in}. The set of patrol points of any robot in R contains only unvisited ones, $T_i = T_i - \{j \mid j \in T_i$ and had been visited$\}$, where T_i denotes the patrol points that had been assigned to robot i, though T_i now contains unvisited ones only. The current locations of robots become the starting locations. The robot that has the highest workload can have one of its patrol points auctioned until the performance can no longer improved. Figure 1 shows the procedure of dynamic patrol planning when the other robots join the mission during the patrolling. The proposed approach first finds the robot that has the maximum patrol path and then performs CAD_W and CAD_L. If the objective is improved, the set of patrol points of the robots is updated. Then CAD_W or CAD_L is performed again for the robot with the maximum patrol path. The procedure will continue until the objective can no longer be improved.

The difference in re-auction procedures between CAD_W/CAD_L and DPP_CAD_W/DPP_CAD_L is only the stopping criterion and the origin of the re-auction patrol points. The stopping criterion of CAD_W or CAD_L is when R_{win} is no longer the robot with the maximum patrol path. However, the procedure of dynamic patrol planning, which is performed when other robots participate in the patrol task, will be stopped when the objective can no longer improved. So CAD_W or CAD_L has the R_{win} that wins the auctioned point during the auction, as the R_{max} all of the time. DPP_CAD_W or DPP_CAD_L, however, needs to find R_{max} in each iteration, in order to choose the patrol point for re-auction from R_{max}.

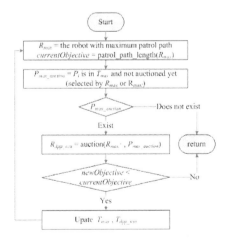

Fig. 1. Procedure of dynamic patrol planning when other robots participate in the patrol mission

4 Simulations and Discussion

In order to evaluate the performance of the proposed dynamic patrol planning algorithm, 10 different maps, 900 patrol points ($|P| = 900$), and 20 robots ($|R| = 20$) are used as the test data. There are two kinds of status during the patrol.

- Decreasing: one robot has malfunctioned and leaves the patrol ($|R|$ decreases by 1 and becomes 19)
- Increasing: an additional robot joins the task during the patrol ($|R|$ increases by 1 and becomes 20)

The current locations of robots and patrol points are distributed on the map according to three different modes so that the simulated environment can be more similar to the real world, as follows:

- Uniform-Uniform (RUPU): the current locations of robots and the locations of the patrol points are uniformly distributed on the map.
- Gather-Uniform (RGPU): the current locations of robots are gathered in a block on the map, but the locations of the patrol points are uniformly distributed on the map.
- Uniform-Gather (RUPG): the current locations of robots are uniformly distributed on the map, but the locations of the patrol points are gathered in a block on the map.

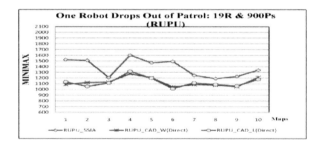

Fig. 2. Comparison of SSIA, DPP_CAD_W, & DPP_CAD_L for one robot dropping out, with uniformly distributed current locations for robots and locations for the patrol points (RUPU)

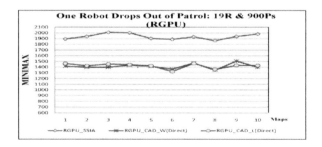

Fig. 3. Comparison of SSIA, DPP_CAD_W, & DPP_CAD_L for one robot dropping out, with gathered current locations for robots, but uniformly distributed locations for the patrol points (RGPU)

Comparison was done among Sequential Single-Item Auction (SSIA), Cooperative Auction Directed by Winner (CAD_W), and Cooperative Auction Directed by Loser (CAD_L). MINIMAX (minimize the maximum patrol path of robots) is the major objective for comparison.

Simulations first consider there are initially 20 robots on duty, but one is out of order and drops out of the task during the patrol. Then, one robot is added during the patrol. The results of simulations for DPP_CAD_W and DPP_CAD_L are shown in Figures 2~7. Figures 2~4 consider a situation in which one robot drops out of the patrol task and Figures 5~7 are the simulation results of adding one robot during the patrol. Brown lines with diamonds denote the SSIA approach, purple lines with asterisks are for the proposed DPP_CAL_W, and green lines with circles are for the proposed DPP_CAL_L in Figures 2~7. Figures 2~7 clearly show that SSIA always has the worst performance for any kind of distribution of the current locations for robots or the locations of the patrol points when other robots quit or join the task during the patrol. RUPG still has the best performance and RGPU still has the worst performance among these distribution patterns. DPP_CAD_L needs more computation time than DPP_CAD_W. However, the objective of our proposed approaches, DPP_CAD_W and DPP_CAD_L, do not show much difference for dynamic patrol planning. CAD_W and CAD_L represent DPP_CAD_W and DPP_CAD_L in Figures 2~7, respectively.

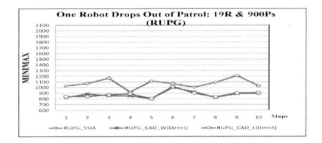

Fig. 4. Comparison of SSIA, DPP_CAD_W, & DPP_CAD_L for one robot dropping out, with uniformly distributed current locations for robots, but gathered locations for the patrol points (RUPG)

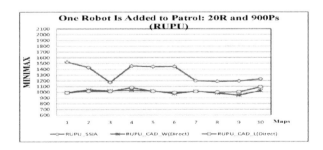

Fig. 5. Comparison of SSIA, DPP_CAD_W, & DPP_CAD_L for one robot being added in, with uniformly distributed current locations for robots and locations for the patrol points (RUPU)

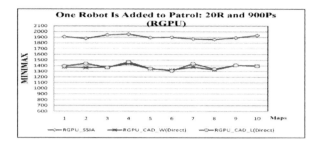

Fig. 6. Comparison of SSIA, DPP_CAD_W, & DPP_CAD_L for one robot being added in, with gathered current locations for robots, but uniformly distributed locations for the patrol points (RGPU)

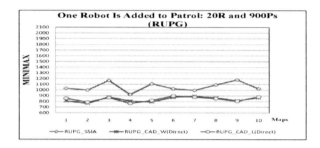

Fig. 7. Comparison of SSIA, DPP_CAD_W, & DPP_CAD_L for one robot being added in, with uniformly distributed current locations for robots, but gathered locations for the patrol points (RUPG)

5   Conclusion

In this article, an auction-based method for the task allocation within a multi-robot patrol system was presented. The robots operate in a structured environment where the costs to the patrol points for each robot are obtainable. Tasks are locations in the map that have to be visited by the robots. Unexpected situations may prevent a robot from being able to complete its allocated tasks. Therefore, tasks not yet achieved by the malfunctioning robots are rebid for by the surviving peers. This provides an opportunity to facilitate the allocation of the remaining tasks and to reduce the overall task completion time. The proposed cooperative auction system which considers re-auction and is based on the performance of team work, can further improve cooperation within a multi-robot system and provide a solution for adapting to constantly changing conditions, where unexpected loss or reinforcement in the number of robots might occur during the patrol.

The large scale of the experimental results of simulation using the proposed task allocation mechanism for multiple robots has shown the performance of the algorithm in terms of efficiency. The simulations indicate that the proposed dynamic patrol planning based on the cooperative auction system has a shorter maximum patrol path for a multi-robot system. The proposed system performs significantly better than that of Sequential Single-Item Auction (SSIA). Moreover, the proposed algorithm has the potential to be scaled to a large number of robots and can respond to the variations of a multi-robot system during the patrol.

References

1. Ugur, A.: Path planning on a cuboid using genetic algorithms. Information Sciences 178, 3275–3287 (2008)
2. Ulas, B., Kemal, L.: Strategy creation, decomposition and distribution in particle navigation. Information Science 177, 755–770 (2007)
3. Tovey, C., Lagoudakis, M.G., Jain, S., Koenig, S.: The Generation of Bidding Rules for Auction-Based Robot Coordination. Multi-Robot Systems from Swarms to Intelligent Automata III, 3–14 (2005)
4. Yang, D., Chen, J., Naofumi, M., Yuzo, Y.: Multi-robot Path Planning Based on Cooperative Co-evolution and Adaptive CGA. In: The IEEE/WIC/ACM International Conference on Intelligent Agent Technology. IEEE Press, Los Alamitos (2006)
5. Guo, G., Zhang, M.: A novel approach for multi-agent-based intelligent manufacturing system. Information Science 179, 3079–3090 (2009)
6. Lagoudakis, M.G., Markakis, E., Kempe, D., Keskinocak, P., Kleywegt, A., Koenig, S., Tovey, C., Meyerson, A., Jain, S.: Auction-Based Multi-Robot Routing. Robotics: Science and Systems (2005)
7. Luiz, S.M.F., Macau, E.E.N.: Patrol Mobile Robots and Chaotic Trajectories. Mathematical Problems in Engineering. Hindawi Publishing Corporation (2007)
8. Mbaitiga, Z.: Intelligent OkikoSenPBX1 Security Patrol Robot via Network and Map-Based Route Planning. Journal of Computer Science 5(1), 79–85 (2009)
9. Koenig, S., Tovey, C., Lagoudakis, M., Markakis, V., Kempe, D., Keskinocak, P., Kleywegt, A., Meyerson, A., Jain, S.: The Power of Sequential Single-Item Auctions for Agent Coordination. American Association for Artificial Intelligence (2006), http://www.aaai.org
10. Hwang, K.S., Lin, J.L., Huang, H.L.: Cooperative Patrol Planning of Multi-Robot Systems by a Competitive Auction System. In: ICCAS-SICE 2009, Fukuoka, Japan (2009)

Vision-Based Robot Manipulator Design

Ching-Chang Wong, Yi-Jiun Shen, Chih-Cheng Liu,
Meng-Tzu Huang, Yu-Ren Huange, and Chen-Yuo Yang

Department of Electrical Engineering, Tamkang University,
Danshui District, New Taipei City, 25137, Taiwan
wong@ee.tku.egu.tw

Abstract. In this paper, a robot with a manipulator, two vision systems, and a three-wheeled omni-directional mobile platform is designed and implemented. A shoulder, an elbow, a wrist, and a movable gripper are designed to let the robot manipulator have 5 degrees of freedom (DOF). The forward kinematics and inverse kinematics are applied to control the manipulator. The forward kinematics is obtained by the D-H (Denavit-Hartenberg) coordinate. The D-H matrix is a coordinate transformed matrix from one coordinate frame to the next one. The inverse kinematics is constructed by the geometry. Two vision systems are applied to search the object and control the manipulator to pick up it. A static camera is used to construct one vision system to get the environment information. An eye-in-hand camera is used to construct the other vision system to track a specific object. Some experimental results are described. The robot manipulator is controlled to draw a picture by using the direct kinematics and geometric inverse kinematics.

Keywords: Mobile Robot, Robot manipulator, Vision System, Kinematics.

1 Introduction

As technology advances, industries are gradually entering the era of fully automated. For example in industry, the industrial robots instead of humans are used to do the repetitive jobs that are high-risk, over sophisticated, too heavy, and long time. In the future life, because of low birth rate and aging are obvious gradually, people hope robots to take care of young persons, offer entertainment, and help elderly or disabled people [1]. The robot manipulators have been applied extensively. The smallest one is applied to assist the minimally invasive surgery and the biggest one is applied to explore the outer space. In this paper, a robot with a manipulator and two vision systems is design and implemented. A 5-DOF robot manipulator is designed to let it has a shoulder, an elbow, a wrist, and a movable gripper. Two vision systems are designed so that the robot can search objects and use the manipulator to pick up them.

The rest of this paper is organized as follows: In Section 2, the system description of the robot is described. In Section 3, the analytic forward and inverse kinematics of the robot manipulator is described. In Section 4, two vision systems are described. In Section 5, some experimental results are presented, where the robot manipulator is controlled to draw a picture by using the forward kinematics and geometric inverse kinematics. Finally, some conclusions are made in Section 6.

T.-H.S. Li et al. (Eds.): FIRA 2011, CCIS 212, pp. 124–131, 2011.

2   System Description of the Robot

The mechanical design of the robot with a three-wheeled omni-directional mobile platform is shown in Fig. 1. The robot's body is made by aluminum. Only one manipulator is mounted in the top body of the robot. The picture of the implemented robot is shown in Fig. 2, where the middle picture is the appearance of the robot. The upper left picture and the upper right picture are the enlarged pictures of the robot manipulator [2]. Eight servo motors with a high torque are used to construct this manipulator. The lower left picture is the enlarged picture of the mobile platform. The lower right picture shows the configuration of three DC motors and three omni-directional wheels, where the gear ratio is used to increase the torque.

The system architecture of the robot is shown in Fig. 3. There are three process cores: a notebook, a Nios II development board, and a CompactRIO single-board. The notebook is used to analyze the image data captured by two cameras, give the move direction to the three wheels, and determine the robot arm's motions. The Nios II development board is used to execute the motion commands of eight servo motors. The CompactRIO single-board is used to determine how many turns are needed to move for the three wheels. The specification of the robot is summarized in Table 1.

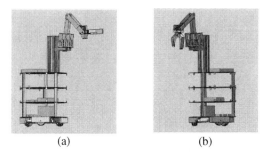

(a)   (b)

Fig. 1. Mechanical design of the robot

Fig. 2. Pictures of the implemented robot and their description

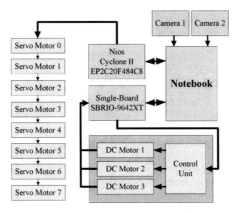

Fig. 3. System architecture of the robot

Table 1. Specification of the robot

Items	Specification	Annotation
Weight	50 kg	Batteries included
Height	135 cm	
Material of structure	TF3030 aluminum	
Shoulder's motor	4	EX-106 (84 kgf.cm)
Elbow's motor	1	RX-64 (64.4 kgf.cm)
Wrist's motor	2	RX-64
Gripper's motor	1	RX-64
Actuator	3	DC motor
Electronic	Nios II EP2C20484C8	
Electronic	Single-Board SBRIO9642XT	
Cinema	320x180 (pixel)	
Batteries	11.1v, 2100mA	Lithium rechargeable battery
Batteries	14.8v, 2100mA	Lithium rechargeable battery

3   Kinematics for Manipulator

As shown in Fig. 4(a), a 5-DOF robot manipulator [3] is designed to let it has a shoulder (1-DOF), an elbow (1-DOF), a wrist (2-DOF), and a movable gripper (1-DOF). The servo motors named RX-64 with a lower torque (64.4 kgf.cm) are used to construct the elbow, wrist and gripper of the robot manipulator. The servo motors named EX-106 with a higher torque (84 kgf.cm) are used to construct the base of the robot manipulator and its shoulder so that the robot manipulator could lift up a heavier object.

3.1   Forward Kinematics

The coordinate space of the manipulator is shown in Fig. 4(b), where l_i ($i=1,2,\ldots,5$) is the length of the i-th link. The objective of forward kinematics analysis [4] is to determine the Cartesian coordinate position (x, y, z) of the end-effector of the manipulator by the joint space $(\theta_1,\theta_2,...,\theta_5)$ of each axel. In this paper, the method of forward kinematics is the Denavit-Hartenberg (D-H) convention, which builds a coordinate as a D-H matrix. As shown in Table 2, there are four parameters in the D-H matrix, where a_i is the distance from Z_{i-1} to Z_i measured along X_i, α_i is the angle from Z_{i-1} to Z_i measured along X_i, d_i is the distance from X_{i-1} to X_i measured along Z_{i-1}, and θ_i is the angle from X_{i-1} to X_i measured about Z_{i-1}. The link coordinate transformation matrices are obtained by

$$A_1^0 = \begin{bmatrix} 1 & 0 & 0 & 0 \\ 0 & 1 & 0 & 0 \\ 0 & 0 & 1 & l_0 \\ 0 & 0 & 0 & 1 \end{bmatrix} \quad A_2^1 = \begin{bmatrix} c\theta_1 & 0 & -s\theta_1 & 0 \\ s\theta_1 & 0 & c\theta_1 & 0 \\ 0 & -1 & 0 & l_1 \\ 0 & 0 & 0 & 1 \end{bmatrix} \quad A_3^2 = \begin{bmatrix} c\theta_2 & -s\theta_2 & 0 & -l_2 c\theta_2 \\ s\theta_2 & c\theta_2 & 0 & l_2 s\theta_2 \\ 0 & 0 & 1 & l_2 \\ 0 & 0 & 0 & 1 \end{bmatrix} \quad (1)$$

and

$$A_4^3 = \begin{bmatrix} c\theta_3 & -s\theta_3 & 0 & -l_3 c\theta_3 \\ s\theta_3 & c\theta_3 & 0 & l_3 s\theta_3 \\ 0 & 0 & 1 & l_3 \\ 0 & 0 & 0 & 1 \end{bmatrix} \quad A_5^4 = \begin{bmatrix} c\theta_4 & 0 & s\theta_4 & 0 \\ s\theta_4 & 0 & -c\theta_4 & 0 \\ 0 & 1 & 1 & l_4 \\ 0 & 0 & 0 & 1 \end{bmatrix} \quad A_6^5 = \begin{bmatrix} c\theta_5 & -s\theta_5 & 0 & -l_4 c\theta_5 \\ s\theta_5 & c\theta_5 & 0 & l_4 s\theta_5 \\ 0 & 0 & 1 & l_4 \\ 0 & 0 & 0 & 1 \end{bmatrix} \quad (2)$$

where $c\theta \equiv \cos\theta$ and $s\theta \equiv \sin\theta$.

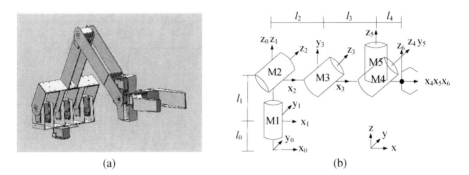

(a)   (b)

Fig. 4. (a) Description of the robot manipulator. (b) Frame design of robot manipulator.

Table 2. D-H parameters of the 5-DOF robot manipulator

Link	a_i	α_i	d_i	θ_i
0	0	0	l_0	0
1	0	-90^0	l_1	θ_1
2	l_2	0	0	θ_2
3	l_3	0	0	θ_3
4	0	90^0	0	θ_4
5	l_4	0	0	θ_5

3.2 Inverse Kinematics

The robot manipulator is given the desired position of end-effector, so the robot manipulator uses the method of inverse kinematics to compute the rotated angle of each joint [5]. In the paper, a geometric approach is used to solve the inverse kinematics problem of 5-DOF robot manipulator with rotary joint. The gripper is the end-effector of the robot manipulator. In order to simplify the calculation, the moving trajectory of the robot manipulator is limited. So, the robot manipulator only moves forward to pick up the object. The wrist can parallel the ground to hold the pen and be perpendicular to the ground to pick up the ball. Therefore, the angle of Motor 4 is obtained. The position of Motor 4 is acquired by the fixed to the gripper distance. The Motor 2, Motor 3 become a two-link mechanism that is shown in Fig. 5(b). When the robot manipulator moves to a new position, the angle of the Motor 1 can be obtained from the position of Motor 4 by the geometry that is shown in Fig. 5(a). The angle of Motor 5 is opposite to Motor 1. Then, the angles of Motor 3 and Motor 4 can be determined by the geometry [6].

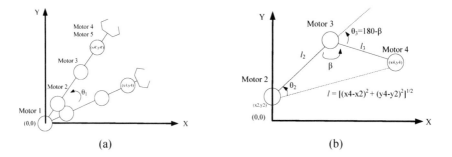

(a)   (b)

Fig. 5. The diagram of the robot manipulator: (a) a vertical view, (b) two-link mechanism

4 Visual Servoing System

The task of the robot is to pick up a circular ball into the box which has the same color with the ball. The main process is to search the ball on the table, move the manipulator to the top of the ball, and use the gripper to pick up the ball [8]. The size of the table is 100×35×100 (cm). In this paper, two vision systems are applied to solve this problem. As shown in Fig. 6, one camera mounts on the robot's gripper that is called an eye-in-hand camera [7]. As shown in Fig. 7, the other camera mounts on the extra bracket that is called a static camera. They are described as follows.

The architecture of the first vision system has six parts and is shown in Fig. 6. The first part is to decide which kind of the color of the ball is picked up first. There are four kinds of the colored balls on the table, so the first order of the ball picked up by the manipulator is based on the distance between the gripper and the colored box. For example, if the gripper is in front of the red box, the manipulator will firstly pick up the red ball. The second part is image preprocessing that the procedure is the process of binary, dilation, erosion, edge, and labeling. The third part is to calculate object position that means calculate the gravity position of the ball. The fourth part is strategic control. Let the ball can be in the catching range. In the fifth part, the robot uses inverse kinematics to move the manipulator to the top of the ball. Finally, the gripper picks up the ball.

The architecture of the second vision system has four parts and is shown in Fig. 7. The task is mainly that uses the static camera to record the relative position of four boxes during the image preprocessing. The strategic control is that after the gripper picks up the ball, the manipulator moves to the front to the same color box by using kinematics, and puts the ball into the box.

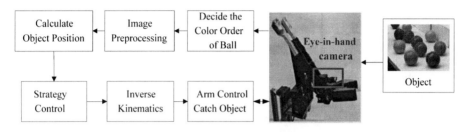

Fig. 6. Architecture of the first visual system

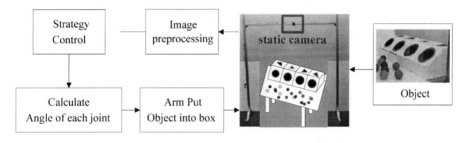

Fig. 7. Architecture of the second visual system

5   Experiments

There are four boxes with different colors of yellow, green, red, and blue on the table. Moreover, each color has three balls on the table. The robot manipulator must to find a ball on the table, pick up it, and put it into a box with the same color of the ball. The experimental results are shown in Fig. 8. The corresponding actions of the manipulator described in Fig.8 (a)~(h) are: (a) use the static camera to decide the color order of ball, (b) use the eye-in-hand to search the ball's position, (c) move the manipulator to the position of the green ball in the catching range, (d) use the gripper to pick up the green ball, (e) hold the green ball, (f) use the static camera to obtain the relative position between the green box and the green ball, (g) move to the top of the green box, and (h) open the gripper to let the ball can fall into the box. In this experiment, the average time is twenty seconds to pick up a ball and put it into the correct box by the robot manipulator.

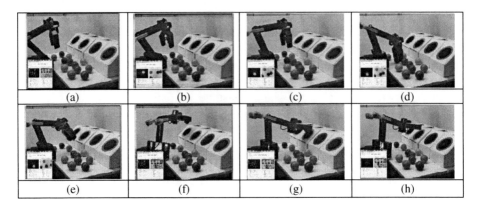

Fig. 8. Description of the robot manipulator picks up a green ball and puts it into a box

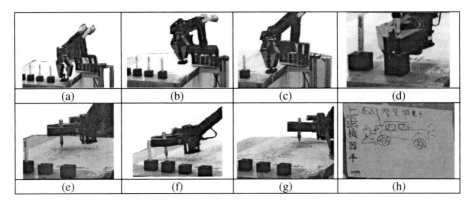

Fig. 9. Description of the robot manipulator stamps the signet and draws a picture

As shown in Fig. 9, the robot manipulator can stamp the signet and draw a picture. The corresponding actions of the manipulator described in Fig.9 (a)~(g) are: (a) move the manipulator to the top of the signet, (b) pick up the signet, (c) stamp the signet, (d) hold the black pen, (e) draw a car shape, (f) hold the red pen to draw a flag, (g) picture the lights of the car. Finally in Fig.9 (f), we can see that a stamp and a car are on the paper.

6 Conclusions

In this paper, a vision-based robot manipulator is designed and implemented. The robot has a manipulator, two vision systems, and a three-wheeled omni-directional mobile chassis. The 5-DOF robot manipulator has a movable gripper. The forward kinematics and geometric inverse kinematics analysis of the robot manipulator are applied to control the manipulator to draw a picture and pick up balls. The robot uses its two vision systems to search and track balls, then the robot controls its manipulator to pick up a ball and put this ball into the box with the same color of the obtained ball.

The experimental results illustrate the robot can combine its vision systems with its manipulator to execute the assigned task efficiently.

Acknowledgment. This research was supported in part by the National Science Council (NSC) of the Republic of China under contract NSC 99-2221-E-032-016-MY2.

References

1. Okada, K., Kojima, M., Sagawa, Y., Ichino, T., Sato, K., Inaba, M.: Vision based behavior verification system of humanoid robot for daily environment tasks. In: 6th IEEE-RAS International Conference on Humanoid Robots, pp. 7–12 (2006)
2. Gueta, L.B., Chiba, R., Arai, T., Ueyama, T., Ota, J.: Design of the end-effector tool attachment for robot arm with multiple reconfigurable goals. In: IEEE International Conference on Automation Science and Engineering, pp. 876–881 (2008)
3. Lee, C.S.G.: Robot Arm Kinematics, Dynamics, and Control. Computer 15, 62–80 (2006)
4. Paul, R.P.: Robot manipulators: mathematics, programming, and control: the computer control of robot manipulators. MIP Press (1981)
5. Guilamo, L., Kuffner, J., Nishiwaki, K., Kagami, S.: Efficient prioritized inverse kinematic solutions for redundant manipulators. In: IEEE/RSJ International Conference on Intelligent Robot and Systems, pp. 3921–3926 (2002)
6. Gou, Z., Sun, Y., Yu, H.: Inverse kinematics equation of 6-DOF robot based on geometry projection and simulation. In: International Conference on Computer, Mechatronics, Control and Electronic Engineering (CMCE), vol. 2, pp. 125–128 (2010)
7. Flandin, G., Chaumette, F., Marchand, E.: Eye-in-hand/eye-to-hand cooperation for visual servoing. In: IEEE International Conference on Robotics and Automation, vol. 3, pp. 2741–2746 (2002)
8. Allen, P.K., Yoshimi, B., Timcenko, A.: Real-time visual servoing. In: IEEE International Conference on Robotics and Automation, pp. 851–856 (2002)

Planar Robot Position and Orientation Measurement Using a Monocular Vision

Yin-Tien Wang[1], Kuo-Wei Chen[1], Po-Hsin Li[1], and Chen-Tung Chi[2]

[1] Mechanical and Electro-Mechanical Eng., Tamkang Univ., New Taipei City, Taiwan
ytwang@mail.tku.edu.tw, 699370218@s99.tku.edu.tw,
495370560@s95.tku.edu.tw
[2] Mechanical Eng., Technology and Science Institute of Northern Taiwan, Taipei, Taiwan
cdchi@tsint.edu.tw

Abstract. A sensor system is developed in the paper to measure the position and orientation of planar robots. In the sensor system, a monocular vision is integrated with a detection method for abstracting the scale- and orientation-invariant image features. Instead of using multiple cameras, a monocular vision is utilized as the only sensing device to reduce the computation cost. The scale- and orientation-invariant method is employed to guarantee a robust detection and description of features abstracted from an image. Experiment is carried out on a free-moving monocular camera to verify the performances of the proposed system.

Keywords: Scale- and orientation-invariant feature, Monocular vision, Robot pose estimation.

1 Introduction

The usage of multiple simple and small-size robots has more advantages than that of single complex and big-size robot in some applications [1-2], for example, manipulation of a large object in a restricted space or clearing the floor of a room. However, simple robots may neither have enough computation power for trajectory planning nor been equipped with sensor to explore the environment. These simple robots are usually supervised and controlled by a system which has sensory and strategy-planning capability. This paper presents a robot pose estimation system to measure the position and orientation of simple and small-size planar robots. A monocular vision is integrated with an image detection and description method in the sensory system to supervise the robots. The monocular camera is utilized as the only sensing device for the estimation system. It is obvious that using monocular vision is more computational efficient in image processing than using multiple camera. Meanwhile, the scale- and orientation-invariant method is employed to guarantee a robust detection and description of features abstracted from an image. These robust features are utilized to construct a sparse map for robot localization as well as to provide apparent point marks on the moving planar robots for efficient estimation of these robots.

2 Monocular Vision Sensor

The proposed sensor system includes a monocular vision and a robust image feature detector. The monocular vision is utilized as the only sensing device. The feature

T.-H.S. Li et al. (Eds.): FIRA 2011, CCIS 212, pp. 132–139, 2011.

detector is based on the concept of Speeded-Up Robust Features (SURF) developed by Bay et al. [3]. The monocular vision model and the SURF method are described in detail in the following subsections.

2.1 Visual Sensor Model

Two coordinate systems, the camera frame $\{C\}$ and world frame $\{W\}$, are set as shown in Fig. 1. The perspective projection method [4] is employed to model the transformation from 3D space coordinate system to 2D image plane. For ith image feature observed by the monocular vision, the measurement is expressed as

$$\begin{bmatrix} I_{ix} \\ I_{iy} \end{bmatrix} = \begin{bmatrix} u_0 + f_c \dfrac{h_{ix}^C}{h_{iz}^C} \\ v_0 + f_c \dfrac{h_{iy}^C}{h_{iz}^C} \end{bmatrix} \tag{1}$$

where (I_{ix}, I_{iy}) are the pixel coordinates of ith image feature; f_c is the focal length of the camera denoting the distance from camera center to image plane; (u_0, v_0) is the offset pixel vector from the hardware image plane to pixel image plane; $h_i^C = [h_{ix}^C \quad h_{iy}^C \quad h_{iz}^C]^T$ is the ray vector of the image features in the camera frame. The 3D coordinates of an image feature, as shown in Fig. 1, are given as

$$Y_i = r + R_C^W h_i^C \tag{2}$$

R_C^W is the rotational matrix from world frame to camera frame and can be represented using the elementary rotations [5]. The components of ray vector can be obtained from Eqn. (1) with measured pixel coordinates. Substituting the ray vectors into Eqn. (2), we can get the measurement of features in world frame in terms of pixel coordinates of the image feature. An alternative method to determine the ray vector of the monocular camera is by using a state estimator, for example, extended Kalman filter (EKF) [6-7].

2.2 Speeded Up Robust Features (SURF)

The basic concept of scale- and orientation-invariant method is to detect image features by investigating the determinant of Hessian matrix H in scale space [8]. In order to speed up the detection of image features, Bay et al. [3] utilize integral images and box filters to process on the image instead of calculating the Hessian matrix, and then the determinant of Hessian matrix is approximated by

$$\det(H)_{approx.} = D_{xx}D_{yy} - (wD_{xy})^2 \tag{3}$$

where D_{ij} are the images filtered by the corresponding box filters; w is a weight constant. The interest points or features are extracted by examining the extreme value of determinant of Hessian matrix. Furthermore, the unique properties of the extracted SURF are described by using a 64-dimensional description vector [9].

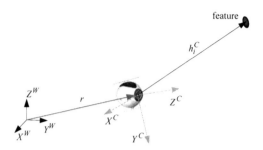

Fig. 1. Monocular vision system

3   Robot Pose Measurement

The camera model described in Subsection 2.1 is utilized to measure the pose of planar robots. One special case is assumed in this section, more general cases can be derived using similar concepts. In this special case, the height of the robot and the width of the field L are known.

3.1   Focus Length

The focus length of the camera is determined by the following steps. Firstly, the camera is located in center-front of the field as shown in Fig. 2. The axes X^C and Y^C of the camera frame are aligned along with X^W and $-Y^W$, respectively. The height from the camera to the ground is N and the horizontal distance to the filed is M. Secondly, the height N and the distance M are adjusted and an angle $-\theta$ along with X^C is rotated. The purpose is to align the image of the line O^WB along with the bottom of the image plane. Finally, one image is captured at this camera location to calculate the focus

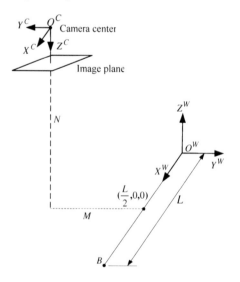

Fig. 2. Monocular vision is located at center-front of the field

length. As shown in Fig. 2, the straight line from the origin O^C of the camera frame to point B in the filed can be expressed as

$$\overline{O^C B} = \sqrt{(u_0 t_1)^2 + (v_0 t_1)^2 + (f_c t_1)^2} = \sqrt{\left(\frac{L}{2}\right)^2 + (-M)^2 + N^2} \tag{4}$$

where $t_1 = L/2u_0$ is the parameter of straight line $O^C B$. Therefore, the camera focus length f_c can be obtained as

$$f_c = \frac{\sqrt{\left(\frac{L}{2}\right)^2 + M^2 + N^2 - (u_0 t_1)^2 - (v_0 t_1)^2}}{t_1} \tag{5}$$

3.2 Ray Vector of Feature Point

As mentioned in Eqn. (2), any space point can be expressed in world frame as $[Y_x, Y_y, Y_z]^T$ or as a ray vector $[h_x^C, h_y^C, h_z^C]^T$ in camera frame. For the camera location described in Subsection 3.1, any space point represented by Eqn. (2) can be rearranged to be a homogenous transformation from world frame to camera frame. The homogenous transformation is defined as

$$\begin{bmatrix} h_x^C \\ h_y^C \\ h_z^C \\ 1 \end{bmatrix} = \begin{bmatrix} 1 & 0 & 0 & -u_0 t_2 \\ 0 & -\cos\theta & -\sin\theta & v_0 t_2 \\ 0 & \sin\theta & -\cos\theta & f_c t_2 \\ 0 & 0 & 0 & 1 \end{bmatrix} \begin{bmatrix} Y_x \\ Y_y \\ Y_z \\ 1 \end{bmatrix} \tag{6}$$

where $t_2 = L/2u_0$ is the parameter of the straight line from the origin O^C of the camera frame to the origin O^W of the world frame. Furthermore, the camera rotation angle θ, as shown in Fig. 3, is calculated as

$$\angle\theta = \theta_1 + \theta_2 = \tan^{-1}\left(\frac{v_0}{f_c}\right) + \tan^{-1}\left(\frac{M}{N}\right) \tag{7}$$

If ith feature point is expressed as Y_i in world frame and h_i^C in camera frame, as shown in Fig. 3, the projection of this feature point on the image plane is P_i^C and expressed in camera frame as

$$P_x^C = (Y_{ix} - u_0 t_2) t_3 \tag{8}$$

$$P_y^C = (-Y_{iy} \cos\theta - Y_{iz} \sin\theta + v_0 t_2) t_3 \tag{9}$$

$$P_z^C = (Y_{iy} \sin\theta - Y_{iz} \cos\theta + f_c t_2) t_3 = f_c \tag{10}$$

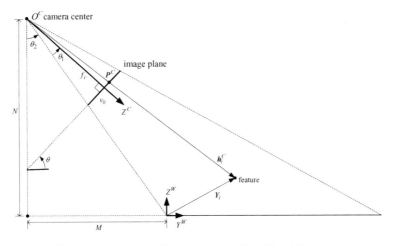

Fig. 3. Ray vector seen from the monocular vision (side view)

where t_3 is the parameter of the straight line from the origin of the camera frame to the feature point. From Eqn. (10), the straight line parameter is determined as

$$t_3 = \frac{f_c}{f_c t_2 + Y_{iy} \sin \theta - Y_{iz} \cos \theta} \tag{11}$$

Substitute Eqn. (11) into Eqns. (8)-(9) to determine the coordinate of $\boldsymbol{P}^C$. From Eqn. (1), the pixel coordinate of ith feature point can be expressed as

$$I_{ix} = \frac{f_c(Y_{ix} - u_0 t_2)}{f_c t_2 + Y_{iy} \sin \theta - Y_{iz} \cos \theta} + u_0 \tag{12}$$

$$I_y = \frac{f_c(-Y_{iy} \cos \theta + v_0 t_2 - Y_{iz} \sin \theta)}{f_c t_2 + Y_{iy} \sin \theta - Y_{iz} \cos \theta} + v_0 \tag{13}$$

Note that this feature point is located on top of the robot and the height Y_{iz} of the planar robot is given. Therefore, the coordinate of ith feature point in world frame can be obtained by solving Eqns. (12)-(13) to get

$$Y_{iy} = \frac{f_c t_2(-I_y + 2v_0) + Y_{iz}((I_y - v_0)\cos \theta - f_c \sin \theta)}{(I_y - v_0)\sin \theta + f_c \cos \theta} \tag{14}$$

$$Y_{ix} = \frac{f_c t_2 I_x + Y_{iy}(I_x - u_0)\sin \theta - Y_{iz}(I_x - u_0)\cos \theta}{f_c} \tag{15}$$

The orientation of the robot is calculated as

$$\alpha = \cos^{-1} \frac{\boldsymbol{h}^T \boldsymbol{u}_x^W}{|\boldsymbol{h}|} \tag{16}$$

where $\boldsymbol{h}$ is vector from point h_2 to h_1 as shown in Fig. 4; $\boldsymbol{u}_x^W$ is the unit vector in x-axis of world frame.

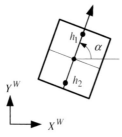

Fig. 4. Orientation of the robot

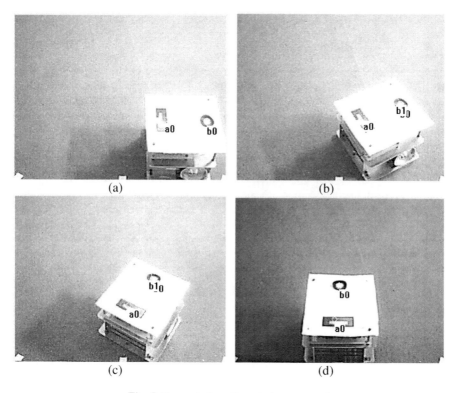

Fig. 5. Four robot positions to be measured

4 Experimental Results

The integrated system is implemented in this paper to demonstrate the proposed algorithm. The experiment is carried out to measure the pose of planar robots. In this example, the state of the monocular camera is estimated with the coordinate ($27cm$, -$46.5cm$, $77cm$). The height of the robot $Y_z=13.5cm$ and the width of the field $L=54cm$ are also given. In this case, the straight line parameter, the camera focal length, and the angle θ are calculated as $t_1=0.17$, $f_c=519.2pixels$, and $\theta=0.77rad$, respectively. The robot pose can be estimated by using Eqns. (14)-(16). The robot pose at four different positions, as shown in Fig. 5, is estimated and the results are listed in Table 1.

Table 1. The real and measured poses of the robots at four different locations

Location	Real poses				measured poses			
	x(cm)	y(cm)	z(cm)	α(degree)	x(cm)	y(cm)	z(cm)	α(degree)
(a)	42.0	8.8		0.0	42.97	7.97		4.4
(b)	37.5	11.0	13.5	30.0	39.42	12.14	13.50	27.5
(c)	31.5	11.2		60.0	32.90	12.80		55.5
(d)	27.0	9.0		90.0	27.02	8.31		84.1

5 Conclusion

This paper presents a monocular visual measuring algorithm to determine the position and orientation of planar robots in the environment. The camera model is utilized to determine the ray vector of the features with given camera location. Meanwhile, the scale- and orientation-invariant image features are detected to represent the moving robot. Experiments have been carried out to verify the performances of the proposed algorithm.

In this paper, we present one special case with specific camera location. More general cases can be derived, if alternative method is utilized to determine the ray vector of the monocular camera. A state estimator, for example, extended Kalman filter (EKF) can be used to estimate the state of the camera and image features.

Acknowledgments. This paper was partially supported by the National Science Council in Taiwan under grant no. NSC98-2815-C-032-011-E to K.W. Chen and NSC99-2221-E-032-064 to Y.T. Wang.

References

1. Mataric, M., Nilsson, M., Simsarian, K.: Cooperative multi-robot box pushing. In: Proceedings of the IEEE/RSJ International Conference on Intelligent Robots and Systems, pp. 556–561 (1995)
2. Chen, Y.-C., Wang, Y.T.: A Generalized Framework of Dynamic Role Assignment for Robot Formation Control. International Journal of Control, Automation, and Systems 8, 1288–1295 (2010)
3. Bay, H., Ess, A., Tuytelaars, T., Van Gool, L.: SURF: speeded up robust features. Computer Vision and Image Understanding 110, 346–359 (2008)
4. Hutchinson, S., Hager, G.D., Corke, P.I.: A tutorial on visual servo control. IEEE Transactions on Robotics and Automation 12, 651–670 (1996)

5. Sciavicco, L., Siciliano, B.: Modelling and Control of Robot Manipulators. McGraw-Hill, New York (1996)
6. Davison, A.J., Reid, I.D., Molton, N.D., Stasse, O.: MonoSLAM Real Time Single Camera SLAM. IEEE Transactions on Pattern Analysis and Machine Intelligence 29, 1052–1067 (2007)
7. Wang, Y.T., Lin, M.C., Ju, R.C.: Visual SLAM and Moving Object Detection for a Small-size Humanoid Robot. International Journal of Advanced Robotic Systems 7, 133–138 (2010)
8. Lindeberg, T.: Feature detection with automatic scale selection. International Journal of Computer Vision 30, 79–116 (1998)
9. Lowe, D.G.: Distinctive Image Features from Scale-Invariant Keypoints. International Journal of Computer Vision 60, 91–110 (2004)

Robot Pose and Velocity Estimation Using a Binocular Vision

Yin-Tien Wang[1], Shi-Hao Wang[1], Ying-Chieh Feng[1], and Jin-Yi Lin[2]

[1] Mechanical and Electro-Mechanical Eng., Tamkang Univ., New Taipei City, Taiwan
ytwang@mail.tku.edu.tw, 496370809@s96.tku.edu.tw,
696372076@s96.tku.edu.tw
[2] Mechanical Eng., Technology and Science Institute of Northern Taiwan, Taipei, Taiwan
jylin@tsint.edu.tw

Abstract. A robot state estimation algorithm based on the vision feedback is proposed in the paper. The algorithm consists of an image feature detector and an extended Kalman filter (EKF) based estimator. The detected image features are scale-invariant and provide a robust representation of moving objects and static landmarks in the environment. The recursive EKF-based estimator is utilized to determine the pose and velocity of moving robots. Experiments are carried out on a hand-held binocular camera to verify the performances of the proposed state estimation algorithm. The results show that the integration of the image feature detector and the state estimator is efficient in highly dynamic environments.

Keywords: Robot State Estimation, Extended Kalman Filter (EKF), Binocular Vision.

1 Introduction

In visual surveillance system, the cameras are usually fixed in space. These cameras might have the capability of directional and zoom (pan-tilt-zoom, PTZ) control, but they lost the mobility in the environment. In this paper, we present a visual surveillance system which is free to move in space. This system is realized by using a free-moving binocular vision which is integrated with a state estimation algorithm. The proposed state estimation algorithm consists of a scale-invariant image feature detector as well as a recursive extended Kalman filter (EKF) to determine the pose and velocity of mobile robots in the environment. This binocular vision may be installed on a mobile robot or any kind of carriers to supervise and control a group of small-size working robots. This paper encompasses a detailed illustration of the components of binocular visual state estimator as well as the experimental results of the implementation of the proposal algorithm.

2 Binocular Visual Surveillance System

The visual state estimator consists of a binocular vision model, the Speeded-Up Robust Features (SURF) [1] detector, and an estimator based on extended Kalman

T.-H.S. Li et al. (Eds.): FIRA 2011, CCIS 212, pp. 140–146, 2011.

filter (EKF). The detail description of SURF detector can be found in the literature [1]. The formulation of the EKF-based estimator and the binocular vision model will be described in the following subsection.

2.1 EKF-Based Estimator

State estimation is a target tracking problem. The state sequence of a system at time step k can be expressed as

$$\mathbf{x}_k = f\left(\mathbf{x}_{k-1}, \mathbf{u}_{k-1}, w_{k-1}\right) \tag{1}$$

where $\mathbf{x}_k$ is the state vector; $\mathbf{u}_k$ is the input; w_k is the process noise. The objective of the tracking problem is to recursively estimate the state $\mathbf{x}_k$ of the target according to the measurement $\mathbf{z}_k$ at k,

$$\mathbf{z}_k = g\left(\mathbf{x}_k, v_k\right) \tag{2}$$

where v_k is the measurement noise. In this paper, a hand-held vision sensor is utilized as the only sensing device for the measurement in SLAM system. We treat this hand-held camera as a free-moving robot system with unknown inputs. The states of the system are estimated by solving the target tracking problem using EKF [2-3],

$$\mathbf{x}_{k/k-1} = f(\mathbf{x}_{k-1|k-1}, \mathbf{u}_{k-1}, 0) \tag{3}$$

$$\mathbf{P}_{k|k-1} = A_k \mathbf{P}_{k-1|k-1} A_k^T + W_k \mathbf{Q}_{k-1} W_k^T \tag{4}$$

$$\mathbf{K}_k = \mathbf{P}_{k|k-1} H_k^T (H_k \mathbf{P}_{k|k-1} H_k^T + V_k \mathbf{R}_k V_k^T)^{-1} \tag{5}$$

$$\mathbf{x}_{k/k} = \mathbf{x}_{k/k-1} + \mathbf{K}_k (\mathbf{z}_k - g(\mathbf{x}_{k/k-1}, 0)) \tag{6}$$

$$\mathbf{P}_{k/k} = (I - \mathbf{K}_k H_k) \mathbf{P}_{k/k-1} \tag{7}$$

where $\mathbf{x}_{k|k-1}$ and $\mathbf{x}_{k|k}$ represent the predicted and estimated state vectors, respectively; $\mathbf{K}_k$ is Kalman gain matrix; $\mathbf{P}$ denotes the covariance matrix, respectively; A_k and W_k are the Jacobian matrices of the state equation f with respect to the state vector $\mathbf{x}_k$ and the noise variable w_k, respectively; H_k and V_k are the Jacobian matrices of the measurement g with respect to the state vector $\mathbf{x}_k$ and the noise variable v_k, respectively.

2.2 Binocular Sensor Model

In the recursive state estimation algorithm, a binocular vision, as shown in Fig. 1, is utilized as the only sensing device for the measurement. The vector of measurement $\mathbf{z}_k$ at time $t=k$ is given as

$$\mathbf{z}_k = [\mathbf{z}_{1k}^T \ \mathbf{z}_{2k}^T \ \cdots \ \mathbf{z}_{mk}^T]^T = g\left(\mathbf{x}_k, v_k\right) \tag{8}$$

where m is the number of measurement at time k. The perspective projection method [4] is employed to model the transformation from 3D space coordinate system to 2D image plane. For ith observed image feature, the measurement is

$$\mathbf{z}_{ik} = \begin{bmatrix} I_{ix} \\ I_{iy} \end{bmatrix} = \begin{bmatrix} u_0 + f_C \dfrac{h_{ix}^C}{h_{iz}^C} \\ v_0 + f_C \dfrac{h_{iy}^C}{h_{iz}^C} \end{bmatrix} \qquad for \ i = 1, 2, \cdots, m \qquad (9)$$

where (I_x, I_y) are the pixel coordinates of a feature in the image plane; f_C is the focal length of the camera denoting the distance from the camera center to the image plane; (u_0, v_0) is the offset pixel vector of the pixel image plane; $h_i^C = [h_{ix}^C \ \ h_{iy}^C \ \ h_{iz}^C]^T$ is defined as the ray vector of the image features in the camera frame. The 3D coordinates of ith image feature or landmark in world frame, as shown in Figure 1, is given as

$$Y_i = r + R_C^W h_i^C \qquad (10)$$

where r is the position vector of the camera frame; R_C^W is the rotational matrix [5] from the world frame {W} to the camera frame {C}. For the binocular vision, the pixel coordinates of two corresponded features in left- and right-image planes are given as

$$I_x^L = u_0 + f_C \frac{h_x^L}{h_z^L} ; \ I_y^L = v_0 + f_C \frac{h_y^L}{h_z^L} ; \ I_x^R = u_0 + f_C \frac{h_x^R}{h_z^R} ; \ I_y^R = v_0 + f_C \frac{h_y^R}{h_z^R} \qquad (11)$$

If the concept of standard stereo geometry (SSG), as shown in Fig. 2, is utilized, the ray vector can be expressed as

$$\mathbf{z} = h^C = \begin{bmatrix} \dfrac{L}{2} \dfrac{(I_x^L - u_0) + (I_x^R - u_0)}{I_x^L - I_x^R} & \dfrac{L}{2} \dfrac{(I_y^L - v_0) + (I_y^R - v_0)}{I_x^L - I_x^R} & \dfrac{L f_C}{I_x^L - I_x^R} \end{bmatrix}^T \qquad (12)$$

The ray vector of the image features in the camera frame is utilized as the prediction of measurement. From Eqns. (9)-(10), the ray vector is determined in terms of the state vector in world frame as

$$\hat{\mathbf{z}} = \hat{h}^C = \begin{bmatrix} c\phi_y c\phi_z (Y_x - r_x) + c\phi_y s\phi_z (Y_y - r_y) - s\phi_y (Y_z - r_z) \\ (s\phi_x s\phi_y c\phi_z - c\phi_x s\phi_z)(Y_x - r_x) + (s\phi_x s\phi_y s\phi_z + c\phi_x c\phi_z)(Y_y - r_y) + s\phi_x c\phi_y (Y_z - r_z) \\ (c\phi_x s\phi_y c\phi_z + s\phi_x s\phi_z)(Y_x - r_x) + (c\phi_x s\phi_y s\phi_z - s\phi_x c\phi_z)(Y_y - r_y) + c\phi_x c\phi_y (Y_z - r_z) \end{bmatrix} \qquad (13)$$

where $s\phi = \sin\phi$ and $c\phi = \cos\phi$. ϕ_x, ϕ_y and ϕ_z are the corresponding rotational angles in world frame.

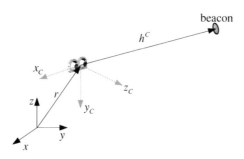

Fig. 1. Binocular vision sensor system

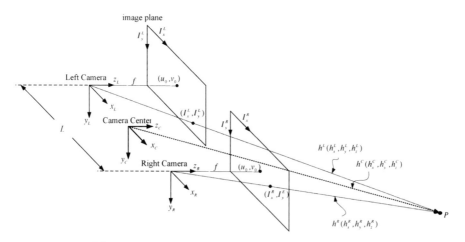

Fig. 2. Perspective projection model of the binocular vision

2.3 Speeded Up Robust Features (SURF)

The basic concept of a scale-invariant method is to detect image features by investigating the determinant of Hessian matrix H in scale space [6]. In order to speed up the detection of image features, Bay et al. [1] utilize integral images and box filters to process on the image instead of calculating the Hessian matrix, and then the determinant of Hessian matrix is approximated by

$$\mathbf{det}(H)_{\text{approx.}} = D_{xx}D_{yy} - (wD_{xy})^2 \tag{14}$$

where D_{ij} are the images filtered by the corresponding box filters; w is a weight constant. The interest points or features are extracted by examining the extreme value of determinant of Hessian matrix. Furthermore, the unique properties of the extracted SURF are described by using a 64-dimensional description vector [7].

2.4 Integrated System

The architecture of the visual surveillance system is depicted in Fig.3. First, the scene is captured by the binocular stereo vision. Apparent features on the small-size robots will be detected and tracked in between two images of the binocular vision. Second, an EKF-based state estimator is utilized to determine the pose and velocity of these robots. Finally, motion control for small-size robots is implemented according to the output of path planning and the visual feedback. Motion command is transmitted to small-size robots by radio frequency (RF) devices like bluetooth. In the integrated system, the surveillance and control system is implemented on a desk-top personal computer (PC). The command decoding and motor driving for small-size robots are executed on an 8051 micro-processor board.

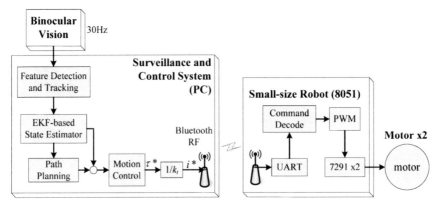

Fig. 3. Architecture of integrated system

3   Experimental Results

Two examples are carried out in this paper to demonstrate the proposed algorithm. These examples are designed for testing the robot motion command in Example 1 and for robot speed estimation in Example 2, respectively.

Example 1. Robot control command testing
In this example, the motion command transmission, decoding, and execution are verified in a real small-size robot system, as shown in Fig. 4. Motion command in

Fig. 4. The small-size mobile robot

Table 1. Binary speed command and the actuated robot speed

Binary command	Robot speed (*cm/sec*)	Binary command	Robot speed (*cm/sec*)
0000	0	1000	29.39±0.38
0001	0	1001	32.88±0.60
0010	0	1010	35.43±0.87
0011	5.85±0.16	1011	37.78±0.67
0100	8.78±0.27	1100	39.98±0.69
0101	11.57±0.27	1101	41.78±0.45
0110	17.04±0.35	1110	42.56±0.68
0111	23.90±0.69	1111	46.91±0.55

four-bit binary format is transmitted from PC to the 8051 board on small-size robots. The speed of the small robot is measured. The results are listed in Table 1. From the table, we can see that if the binary command smaller than 0010, the speed of the robot is zero due to static friction. The maximum speed of the small-size mobile robot is 46.91*cm/sec*. For each of the measurement, 5 samples are taken. The results in the table are averaged measured velocity and the ± variation values indicate the standard deviation of the sampled measured values.

Example 2. Robot speed estimation
The speed of one small-size robot in the environment is determined using the proposed EKF-based state estimator. The computer interface for the binocular state estimator is depicted in Fig. 5. Stereo vision is established using both images captured by the binocular camera. Apparent features on the moving object are detected and tracked in between left- and right-image. The EKF-based state estimator is utilized to determine the robot velocities and the results are listed in Table 2. For each of the estimation, 5 samples are taken. The results in the table are averaged estimated velocity from the EKF-based estimator and the ± variation values indicate the standard deviation of the sampled estimated values.

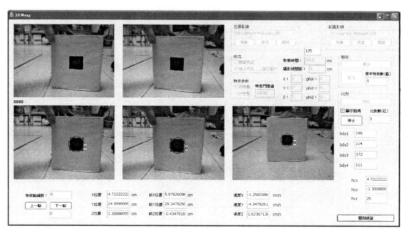

Fig. 5. Computer interface for the state estimator using binocular vision

Table 2. The real and estimated velocities of the robots in three cases

	Real velocity (cm/sec)			Estimated velocity (cm/sec)		
Cases	V_x	V_y	V_z	V_x	V_y	V_z
1	-5.00	0.00	0.00	-4.39±0.48	-0.25±1.11	-0.01±0.08
2	0.00	5.00	0.00	-0.16±0.19	5.30±0.79	-0.34±0.34
3	0.00	0.00	5.00	-0.08±0.17	1.45±0.95	5.42±0.39

4 Conclusion

This paper presents a robot state estimation algorithm based on the vision feedback to determine the pose and velocity of a robot in the environment. The scale- and orientation-invariant image features are detected to represent the moving robot and the static landmarks in the environment. Meanwhile, the EKF-based estimator is utilized to determine the pose and velocity of small-size robots in the environment. Experiments have been carried out on a hand-held binocular camera to verify the performances of the proposed algorithm.

Acknowledgments. This paper was partially supported by the National Science Council in Taiwan under grant no. NSC99-2815-C-032-012-E to S.H. Wang and NSC99-2221-E-032-064 to Y.T. Wang.

References

1. Bay, H., Ess, A., Tuytelaars, T., Van Gool, L.: SURF: speeded up robust features. Computer Vision and Image Understanding 110, 346–359 (2008)
2. Davison, A.J., Reid, I.D., Molton, N.D., Stasse, O.: MonoSLAM Real Time Single Camera SLAM. IEEE Transactions on Pattern Analysis and Machine Intelligence 29, 1052–1067 (2007)
3. Wang, Y.T., Lin, M.C., Ju, R.C.: Visual SLAM and Moving Object Detection for a Small-size Humanoid Robot. International Journal of Advanced Robotic Systems 7, 133–138 (2010)
4. Hutchinson, S., Hager, G.D., Corke, P.I.: A tutorial on visual servo control. IEEE Transactions on Robotics and Automation 12, 651–670 (1996)
5. Sciavicco, L., Siciliano, B.: Modelling and Control of Robot Manipulators. McGraw-Hill, New York (1996)
6. Lindeberg, T.: Feature detection with automatic scale selection. International Journal of Computer Vision 30, 79–116 (1998)
7. Lowe, D.G.: Distinctive Image Features from Scale-Invariant Keypoints. International Journal of Computer Vision 60, 91–110 (2004)

Learning of Facial Gestures Using SVMs

Jacky Baltes, Stela Seo, Chi Tai Cheng, M.C. Lau, and John Anderson

Autonomous Agent Lab,
University of Manitoba,
Winnipeg, Manitoba,
Canada, R3T 2N2
j.baltes@cs.umanitoba.ca
http://www.cs.umanitoba.ca/~jacky

Abstract. This paper describes the implementation of a fast and accurate gesture recognition system. Image sequences are used to train a standard SVM to recognize Yes, No, and Neutral gestures from different users. We show that our system is able to detect facial gestures with more than 80% accuracy from even small input images.

Keywords: Facial Recognition, SVM, Machine Learning.

1 Introduction

In this paper, we are presenting a machine learning approach to user independent facial gesture recognition. The motivation is to learn a set of simple facial gestures that are nevertheless practical and often useful in human robot interaction.

A SVM is used to train the system to detect three different gestures: Yes, No, and Neutral as shown in Fig. 1. We investigate several pre-processing steps as described in 3.2 to avoid overfitting and to speed up the facial gesture detection.

2 Related Work

In recent years, advances in robot hardware and software has led to increased interest in the area of human robot interaction [5]. Many researchers have looked at various aspects of full body gesture recognition for human robot interaction [4].

Many researchers have investigated the use of computer vision in facial gesture recognition for human robot interaction, since vision provides a cheap and ubiquitous, but extremely flexible and powerful sensor [3]. However, whereas recognizing gestures and motions comes naturally to humans and animals, it is very difficult for computers and robots. This is especially true in unstructured environments where the lighting and background can not carefully be controlled [7]. Researchers use mid-level vision features (e.g., the output of region segmentation and matching or objects grouped through optical flow) to detect and classify gestures. These features are then grouped into gestures using finite state machines, particle filters, neural networks, and hidden Markov models.

T.-H.S. Li et al. (Eds.): FIRA 2011, CCIS 212, pp. 147–154, 2011.

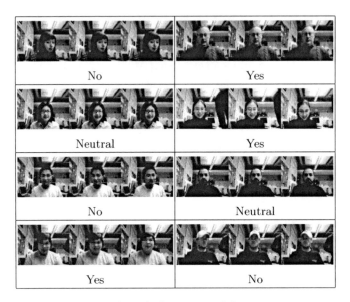

Fig. 1. Sample Sequences of Gestures

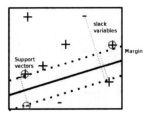

Fig. 2. Linear SVM example showing support vectors, margin, and slack variables

Several researchers have investigated the use of neural networks in recognizing facial gestures [2]. However, these systems are not in common use because their accuracy is not high enough in unstructured environments and without time consuming calibration.

In recent years, support vector machines (SVM) first introduced by Vapnik in 1995 have shown very good performance in a range of supervised classification problems [8]. Support vector machines try to learn a decision surface that maximizes the separation between positive and negative instances of the decision problem. This separation is called the margin and the instances that lie along the margin are called support vectors, which give SVMs their name. An example of a two-dimensional SVM is shown in Fig. 2.

The simplest form of SVM learns linear classifiers which break the training data, which is represented as an n dimensional input vector, into two classes using a $n-1$ dimensional hyperplane. These so-called hard margin Linear SVMs may fail if the training set cannot be separated by a hyperplane.

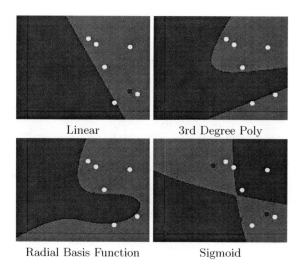

Linear   3rd Degree Poly

Radial Basis Function   Sigmoid

Fig. 3. SVM examples for different decision surfaces. Training data is shown as dots.

A commonly used extension of SVMs are soft-margin SVMs, that do try to both (a) maximize the margin, and (b) minimize the distance of misclassified instances to the boundary of the correct margin boundary. This distance is modelled using so-called slack variables.

SVMs have also been used with more complex decision surfaces such as second, third, or higher order polynomials, radial basis functions, and sigmoid functions. Figure 3 shows a randomly created two-dimensional sample problem. The training data is shown as blue and green dots and the decision surfaces are shown in the corresponding colour.

Other researchers have attempted to use SVMs to learn gestures using support vector machines [6].

3   Design

Implementing a SVM based machine learning system required the implementation of a quadratic programming problem solver, which is non-trivial. Luckily, a high quality open-source library LIBSVM was developed and made available as open source software by Chang, Chih-Chung and Lin, Chih-Jen [1].

LIBSVM has bindings to various programming languages include C,Java, and Python. It provides several optimizations and improvements to the original SVM model. It is fast, reliable, and well documented. We therefore chose to use LIBSVM as the SVM solver in our application. The system described in this paper was implemented in Java and the complete code can be downloaded from our lab website (http://aalab.cs.umanitoba.ca).

3.1   Conversion of Images into Instance Vectors

The images must be converted into real-valued vectors so that they can be used as inputs to an SVM. One issue is that intuitive conversions of the input to a real-valued scalar may result in greatly different scales for the different dimensions of the vector. For example, if we are trying to learn a classifier for predicting a disease then we may want to include body weight as well as amount of iron in the blood stream. However, typical body weight will be around 75 kg, whereas iron will be measured in milligrams. But classification using SVMs is based on the dot product of the instance and the weight vector, which corresponds to the angle between the instance and weight vector. Any change in the iron vector will be overshadowed by changes in the weight vector.

Therefore, scaling of the real-valued input vector is an important step that determines the success of the SVM. This is easily done in images, since the minimum and maximum of the different channels is known. We normalize each colour channel (red, green, or blue) of a pixel or brightness of a pixel into a value between 0 and 1, by dividing it with 255, the maximum value for each channel or brightness. The values of the difference images are normalized by dividing each value with two times the maximum and adding a constant offset of 0.5 to the value for colour or greyscale difference images respectively.

The vector for an instance is created from an image by joining pixels from the top left to the bottom right along each row.

3.2   Pre-processing

In most machine learning applications, too much data is just as bad as too little data. The reason for this is that too much input data often leads to overfitting because features that are not relevant to the target concept may occur in significant patterns by chance. This problem is called the curse of dimensionality.

We therefore investigated several pre-processing methods to reduce the size of the input from the original 240*60*3 dimensional vector. We considered four direct and two difference representation pre-processing steps as described in the following.

Direct Representation. Firstly, we converted the colour images to grey-scale images. Each pixel was replaced by the average brightness of the red, green, and blue channels.

Secondly, we sub-sampled the images by a factor of four, which resulted in 60*15 sized images. Each pixel in the sub-sampled image is the average of four adjacent pixels in the original image.

Difference Representation. We also investigated a difference representation for the input images. In that case, an image was created by subtracting the first image from the second image in the sequence and another image by subtracting the second image from the third image in the sequence. The two images were then combined into an input image. A sample difference image is shown in Fig. 4.

$\Rightarrow$

Fig. 4. Sample difference image

4   Evaluation

We evaluate the performance of the system by creating a database of eight different subjects. Each subject was asked to show three gestures (Yes, No, and Neutral) for approximately 30 seconds each.

It is difficult to recognize gestures using adjacent frames in the video sequence since there is only very small differences between successive frames in a video stream at high frame rates. For example, the difference between images is only 33.30ms at 30 frames per second. We therefore create frame sequences by selecting three images that are spaced approximately 200ms apart and discarding the intermediate frames. This results in three images where the first and the middle and the middle and the end frame are 200ms apart.

Approximately one second after the instruction is given to the subject we start to record video and extract approximately 70 frame sequences from the video randomly. This results in approximately 70 frame sequences for each of three gestures (Yes, No, Neutral) for eight subjects or a total of 1572 frame sequences in our database.

To avoid biasing the sequences, the selection and order of subjects was done at random and all other factors (e.g., time of day) was kept as constant as practical.

The algorithms were trained and their generalization ability tested using a n percent holdout method. That is, n percent of the sequences were selected at random and used for training.

SVM only provide binary decision procedures. We therefore use the following common approach to solving the multi-variate decision problem. Three separate SVMs are trained for Yes, No, and Neutral respectively. For example, the SVM for Yes uses sequences of Yes as positive examples and all other sequences (No, Neutral) as negative examples. Similarly we train the No and Neutral SVMs.

After training, the system classifies all 1572 sequences in the database. We measured the accuracy of the system as the percentage of correctly classified sequences in the database.

An sequence is classified by computing its classification using the Yes, No, and Neutral SVM and by comparing the strength of the classifier in the positive region. The sequence is classified as the gesture associated with the SVM that returns the maximum in the positive region.

4.1   Pre-processing

First, we compare the performance of the SVM using various pre-processing steps described in section 3: (a) colour image 80x60, (b) colour sub-sampled image

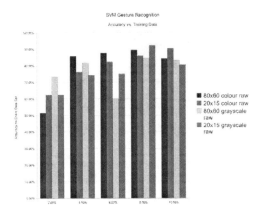

Fig. 5. Accuracy vs. training instances grouped by pre-processing method

20x15, (c) grey-scale 80x60 image, (d) grey-scale sub-sampled image (20x15), (e) difference image 80x60, (f) difference sub-sampled image 20x15.

Figure 5 shows the accuracy of the SVM gesture recognizer with different number of training instances. The influence of the number of training instances is discussed in Subsec 4.2.

The entries for 6%, 8%, and 10% show that there is some variation in the pre-processing methods and that no clear winner emerges. In general the larger image size methods (80x60) seem to perform better than the sub-sampled image methods (20x15), but that difference is not statistically significant. With sufficient training instances, the performance of the four methods described above is above 80%, which is suitable for our application.

4.2   Number of Training Instances

The first characteristic we investigated is the number of training examples needed to train an accurate classifier. Figure 5 shows the accuracy of various pre-processing methods dependent on the number of training examples.

The number of training samples is an important parameter of the algorithm since the run time and storage complexity of an SVM algorithm grows with the square of the number of training samples.

We trained the SVM gesture recognizer using an increasing number of training samples from 2% to 10% of the dataset. This corresponds to 31 to 155 training images out of a total 1572 images in the data set. The accuracy was evaluated using the full 1572 images.

As can be seen, the performance of the system is poor if only two percent of the data set is used as training data, but improves quickly. There is no statistical difference in the performance of the system with 8% (125 images) or 10% (157 images) of training instances.

The effect of increasing the number of training instances on the runtime is shown in Fig. 6.

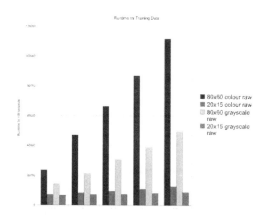

Fig. 6. Runtime vs. training instances grouped by pre-processing method

As can be seen, the runtime grows steeply with increasing the number of training samples for the large image size. Reducing the size of the image greatly reduces the runtime of our algorithm from 1 minute 51 seconds to 8.64 seconds.

4.3 Difference Representation

As can be seen in Fig. 7, the performance of the difference methods is worse than expected. The accuracy on the entire data set is only 40%.

This is due to the fact that as in most robotics applications, calculating the difference between sensor signals greatly increases the suscebitability to noise. So small noisy inputs are amplified by calculating the derivative of the input signal. In this case, the few detected true motion pixels are overshadowed by background noise.

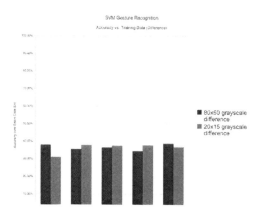

Fig. 7. Accuracy vs. training instances for difference representations

5   Conclusion

We describe a fast and practical SVM learning system for detecting facial gestures indicating Yes, No, or Neural expressions. It is shown that even small images of size 20x15 are suitable to detect the gesture. This means that the system can be combined with a face detector to watch multiple users in a single view.

References

1. Chang, C.-C., Lin, C.-J.: Libsvm: a library for support vector machines. Computer, 1–30 (2001)
2. Hagg, J., Rkl, B., Akan, B., Asplund, L.: Gesture recognition using evolution strategy neural network, pp. 245–248. IEEE, Los Alamitos (2008)
3. Hasanuzzaman, M., Ampornaramveth, V., Bhuiyan, M.A., Shirai, Y., Ueno, H.: Real-time vision-based gesture recognition for human robot interaction. In: 2004 IEEE International Conference on Robotics and Biomimetics, pp. 413–418 (2004)
4. Lee, S.-W.: Automatic gesture recognition for intelligent human-robot interaction. In: 7th International Conference on Automatic Face and Gesture Recognition FGR 2006, pp. 645–650 (2006)
5. Mitra, S., Acharya, T.: Gesture recognition: A survey. IEEE Transactions on Systems Man and Cybernetics Part C Applications and Reviews 37(3), 311–324 (2007)
6. Oshita, M., Matsunaga, T.: Automatic learning of gesture recognition model using som and svm. Advances in Visual Computing, 751–759 (2010)
7. Valibeik, S., Yang, G.-Z.: Segmentation and Tracking for Vision Based Human Robot Interaction. IEEE, Los Alamitos (2008)
8. Vapnik, V.N.: An overview of statistical learning theory. IEEE Transactions on Neural Networks 10(5), 988–999 (1999)

Simulation-Based Analysis and Experimental Verification of Chaotic Circuits

Pao-Lung Chen and Ke-Xin Lin

Dept. of Computer and Communication Engineering,
National Kaohsiung First University of Science and Technology,
No. 2, Jhuoyue Road, Nanzih District, Kaohsiung City 811, Taiwan, R.O.C.
plchen@nkfust.edu.tw

Abstract. This paper presents a simulation-based analysis of chaotic circuits. In the simulation and experiments, the inductor is replaced with one operational amplifier (OP) and other linear components. The negative resistance is realized with two μA741 OPs. At first, we discuss the low loss chaotic circuit. Then, we proposed a lossless chaotic circuit. Finally, we verify those simulations with experimental results. Applications of chaotic circuits to robot will also be described.

Keywords: Chaotic circuits, frequency domain analysis, robot control.

1 Introduction

Chaos is a widely phenomenon in dynamical systems which can be viewed through chaotic circuits. In addition, the chaotic circuits can be applied in many areas of applications such as chaos secure communication, random number generator and robot control. Chua's circuit was proposed by Professor Chou in 1982 [1] due to its easy implementation as well as robustness to parametric mismatches. The study of nonlinear electronic circuits has received great attention in the past decades [2]. However, it required a noncommercial valued inductor for Chou's circuit. Active operational amplifier is used in [3] with loss of energy. In this paper, we presented a simulation based analysis of chaotic circuits. We propose a lossless chaotic circuit. In addition, we also verified it experimental results.

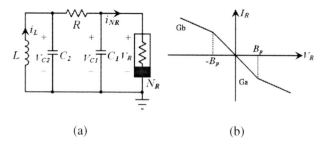

Fig. 1. (a) Chua's chaotic circuit. (b) (I/V) characteristics of nonlinear element.

T.-H.S. Li et al. (Eds.): FIRA 2011, CCIS 212, pp. 155–161, 2011.

A basic third-order chaotic circuit is shown in Fig. 1 (a). It consists of three basic circuit elements. One is the linear inductor L. The second is the two capacitors C_1 and C_2. The thirds are the resistors which includes one linear resistor R and the Chua's diode N_R. The nonlinear circuit of Chua's diode is a critical element which has three segments with negative resistance N_R [4] as shown in Fig. 1 (b).

The circuit equations in Fig. 1 can be represented in the following differential equations [1]:

$$C_1 \frac{dV_{C1}}{dt} = \frac{1}{R}(V_{C2} - V_{C1}) - i_{NR} \tag{1}$$

$$C_2 \frac{dV_{C2}}{dt} = \frac{1}{R}(V_{C1} - V_{C2}) + i_L \tag{2}$$

$$L \frac{di_L}{dt} = -V_{C2} \tag{3}$$

$$i_{NR} = f(V_{C1}) = G_b V_{C1} + \frac{1}{2}(G_a - G_b)(|V_{C1} + B_p| - |V_{C1} - B_p|) \tag{4}$$

The I/V characteristics of i_{NR} is shown in Fig. 1(b). The slopes of the inner and outer regions are G_a and G_b, respectively, while B_p indicates the breakpoints as indicated in equation (4). Fig. 2 is the the implementation of nonlinear resistor N_R. The diodes of D_1 and D_2 are 1N4148. The resistive value of each resistor is $R_1=R_2=220\Omega$, $R_3=1.285K\Omega$, $R_4=R_5=3.3K\Omega$ and $R_6=R_7=47K\Omega$. The operational amplifier is uA741 and $V_{EE} =+9V$, $V_{CC}=-9V$. The value of G_a is -1.072×10^{-3}, G_b is 3×10^{-4} and the B_p is 6.83V [4].

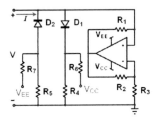

Fig. 2. Implementation of negative resistor N_R

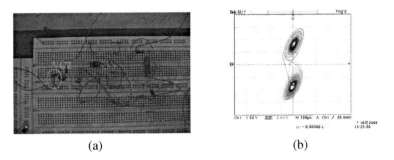

(a)   (b)

Fig. 3. a). Experiment of Chua's circuit b). Measurement of double-scroll(V_{C2} v.s V_{C1})

When L=10mH , C_2=83nF , C_1=4.5nF and R=1.760KΩ, Fig. 3(a) is the experimental verification of Chua's circuit based on discrete components. Fig. 3(b) is the measurement of double scrolls (V_{C2} v.s. V_{C1}).

2 Characteristics of Low Loss Chaotic Circuit

To investigate the synchronization of identical chaotic systems, we would like to fix the parameters of the Chua's circuits as in [5]. The system exhibits a chaotic attractor; specially the double-scroll attractor [6,7]. A low loss chaotic circuit with three μA 741 Ops is shown in Fig. 4. The inductor is replaced with an operational amplifier, resistors and capacitor. The capacitance of C_1 and C_2 are 10nF and 100nF, respectively. Two μA741 Ops and four resistors are used to implement the nonlinear element N_R. The linear inductor has been implemented by one μA741, one capacitor 2.88uF, two 500Ω resistors and two 12.5Ω resistors. We obtain chaotic behavior with double scroll for different value of R in the range of 1390Ω < R < 1471Ω.

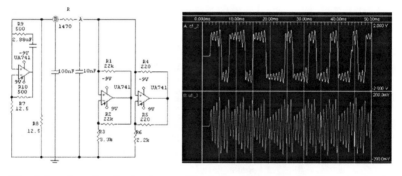

Fig. 4. Low loss chaotic circuits   **Fig. 5.** Simulation out waveform at point A and B

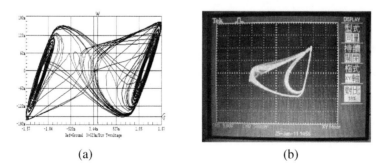

(a)   (b)

Fig. 6. Low lost chaotic circuits for transient analysis of V_{C1} v.s. V_{C2}. (a) Simulation. (b) Experimental result.

Fig. 5 is the simulation of voltage output waveform at point A (V_{C1}) and point B (V_{C2}). Point A is the voltage on capacitance C_1=10nF and point B is voltage across C_2=100nF. The voltage at point A has large swing as compared with point B due to smaller capacitance value. The voltage output swing of point A and point B is inversed with the value of capacitors. The voltage of point A will oscillate in two different states. The X-Y mode simulation of V_{C1} v.s. V_{C2} is shown in Fig. 6(a). Fig. 6(b) is the experimental result. The chaotic circuit is in two oscillation cycles which is called double-scroll. The X-axis is the output voltage of V_{C1} and the Y-axis is the output voltage of V_{C2}.

3   Proposed Chaotic Circuits with Lossless Inductor

We proposed a lossless inductor using μA 741 Ops is shown in Fig. 7. The major difference of Fig.7's realization of inductor as compared with Fig. 4's realization of low lost inductor. The derivation of the equivalent circuit is described in equation (5). In Fig. 7, the V_1 is the input voltage and the i is the input current and of the equivalent inductor. The voltage at the output of operational amplifier is Vo.

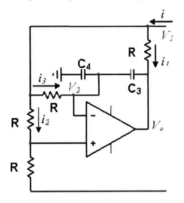

Fig. 7. Equivalent circuit for lossless inductor

$$V_o = \frac{jwC_4R_2 + jwC_3R_2 - 1}{2jwC_3R_2} V_1$$

$$i = \frac{2V_1 - V_o}{R} = \frac{3jwC_3RV_1 - jwC_4RV_1 + V_1}{2jwC_3R^2}$$

$$\Rightarrow Z_{in} = \frac{V_1}{i} = \frac{6w^2C_3^2R^3 - 2w^2C_3C_4R^3 + jw2C_3R^2}{1 + (3wC_3R - wC_3R)^2}$$

$$= \frac{6w^2C_3^2R^3 - 2w^2C_3C_4R^3}{1 + (3wC_3R - wC_4R)^2} + jw\left[\frac{2C_3R^2}{1 + (3wC_3R - wC_4R)^2}\right]$$

$$= R_{eq} + jwL_{eq} \tag{5}$$

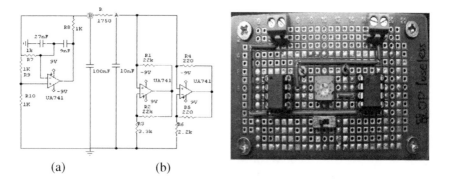

(a)   (b)

Fig. 8. A lossless chaotic circuit. (a). Circuit diagram. (b). Experimental board.

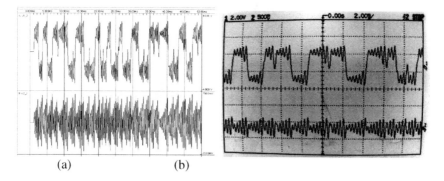

(a)   (b)

Fig. 9. Voltage output waveform of V_{C1} and V_{C2} (a). Simulation (b). Experimental result

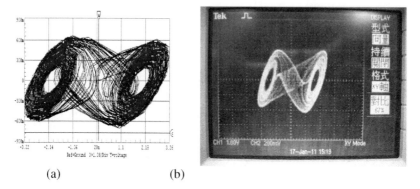

(a)   (b)

Fig. 10. Lossless chaotic circuit for transient analysis of V_{C1} v.s. V_{C2}. (a). Simulation (b). Experimental result.

If we let $R=1K\Omega$ and $C_3=9nF$, $C_4=27nF$, then we can get $L_{eq}=18mH$ by using the equation (5). In Fig. 8, the realization of negative resistance N_R is the same as in Fig. 4. The value of capacitors C_1 and C_2 are 10nF and 100nF, respectively. Fig. 8(b) is the experimental circuit board to implement the lossless chaotic circuit.

Fig. 9(a) is the simulation of voltage output waveform at point A (V_{C1}) and point B (V_{C2}) when resistor R is 1750 Ω. Fig. 9(b) is the measurement from experimental circuit. The amplitude of V_{C1} and V_{C2} has large swing. The X-Y mode simulation of V_{C1} v.s. V_{C2} is shown in Fig. 10(a). Fig. 10(b) is the experimental result. The X-axis is the output voltage of V_{C1} and the Y-axis is the output voltage of V_{C2}.

3.1 Frequency Analysis of Low Lost Chaotic Circuit

Fig. 11 and Fig. 12 are the frequency spectra of V_{C1} and V_{C2} when R is 1750 Ω. The frequency spectra distributions of V_{C1} and V_{C2} are different. The frequency distributions of V_{C1} spread over 5.9KHz. However, the V_{C2} has three main frequency locations on 1KHz, 2KHz and 5KHz.

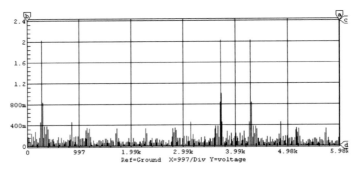

Fig. 11. Frequency spectra of V_{C1}

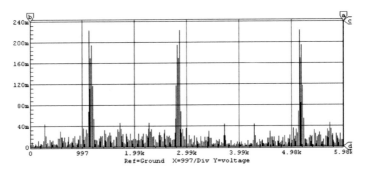

Fig. 12. Frequency spectra of V_{C2}

3.2 Comparison of Low Lost and Lossless Chaotic Circuits

The lossless chaotic circuit has better performance as compared with low lost chaotic circuit. The lossless chaotic circuit has larger amplitude and wider frequency spectrum. Table 1 is the comparison of low lost inductor and lossless inductor.

Table 1. Performance comparison

Type	Low-lost inductor	Lossless inductor
V_{C1}'s amplitude	175mv	600mv
V_{C2}'s amplitude	1.57V	3.26V
Inductor	1OP, 2.88uF, 12.5 Ω	1OP, 27nF, 9nF, 1KΩ
Spectrum	Small range	Large range

3.3 Applications to Robot Control

The lossless chaotic circuit is excellent in secure communication which can be applied in remote robot control. In addition, the chaotic circuit can generate the uniform random number which could be used as a decision mechanism for random walking.

4 Conclusion

In this paper, we present a simulation based analysis of low lost chaotic circuits. We also proposed a lossless chaotic circuit. The lossless chaotic circuit has wider range of chaotic behavior as compared with low lost chaotic circuit. All of the circuits have been simulated and verified by experimental results. The negative resistance is replaced by operational amplifiers and diodes. The inductor is also replaced with operational amplifiers and other discrete components. Finally, we applied the frequency domain analysis for all of the chaotic circuits. The frequency distribution of the chaotic circuit located under 5KHz. The proposed chaotic circuit can be applied in the in the secure communication for robot control.

Acknowledgments. The authors would like to thank the support from NSC-99-2221-E-327-047.

References

1. Matsumoto, T.: A Chaotic Attractor from Chua's Circuit. IEEE Transactions on Circuits and Systems CAS31(12), 1055–1058 (1984)
2. Chua, L.O., Wu, C.W., Huang, A., Zhong, G.-Q.: A universal circuit for studying and generating chaos- part I: Routes to Chaos. IEEE Trans. Circuit Syst. 40(10), 732–744 (1993)
3. μA741A Datasheet, Texas Instruments Inc. (2000), http://www.ti.com
4. Matsumoto, T., Chua, L.O., Komuro, K.: The double scroll. IEEE Trans. Circuit Syst. 32(8), 798–818 (1985)
5. Morgül, Ö.: Inductorless realization of Chua's oscillator. Electronic Letters 31(17), 1403–1404 (1995)
6. Torres, L.A.B., Aguirre, L.A.: Inductorless Chua's circuit. Electronic Letters 36(23), 1915–1916 (2000)
7. Mahesh, M., Cherif, A.: Implementation of the Chua's circuit and its applications. In: Proceedings of the 2002 ASEE Gulf-Southwest Annual Conference (2002)

Codebook Model for Real Time Robot Soccer Recognition: A Comparative Study

Doli Anggia Harahap, Anton Satria Prabuwono, and Azizi Abdullah

Center for Artificial Intelligence Technology (CAIT),
Faculty of Information Science and Technology, Universiti Kebangsaan Malaysia,
43600 UKM Bangi, Selangor D. E, Malaysia
doli.hrp@gmail.com, {antonsatria,azizi}@ftsm.ukm.my

Abstract. Background subtraction is one of several image segmentation techniques. This technique is used in conditions where the background is boring and static, such as in video surveillance. The codebook model is one of the latest and best techniques utilized for background subtraction. Implementing this technique for robotic soccer vision is a good idea. However, the robotic soccer application needs very fast and robust image pre-processing for image segmentation. We slightly modified the codebook algorithm to get the best performance to be implemented in robotic soccer vision. The result of the experiment shows that the performance of the algorithm becomes better.

Keywords: Background subtraction, codebook model, robot soccer, image segmentation.

1 Introduction

Background subtraction is the main idea of image segmentation for detecting moving objects in video surveillance, traffic monitoring and other systems which involves static backgrounds. The main principle is the comparison of the condition in the current image with the background model. This technique is superior for use with video which has a static and boring background. In relation to this, robotic soccer applications also have static backgrounds in the form of the soccer field. Thus, in this paper we have tried to implement the background subtraction technique to overcome this constraint. The idea is to subtract the robots from the soccer field.

An adaptive technique was introduced for estimating the background based on learning which was calculated after every frame processed [1].This algorithm was computationally expensive. Another technique was introduced similar to the one before, and the learning process is based on distance and color but still the computational process was expensive [2]. The simplest technique for background modeling was to assume that intensity values can be modeled by a single distribution [3]. The single-mode model is hard to handle for slight changes of background, like those due to swaying trees, moving curtains or perhaps changes of illumination.

The Mixture of Gaussians (MOG) method [4] was introduced to model more complex non-static backgrounds. Another method based on MOG has been widely proposed, such as utilization of the Bayesian frameworks [5]. MOG has performed

T.-H.S. Li et al. (Eds.): FIRA 2011, CCIS 212, pp. 162–169, 2011.

well, but it has some disadvantages. For a low learning rate, it produces a very wide model. For objects that move slowly, the possibility that they will be absorbed into the background model was high. This problem is described in [6].

The Codebook (CB) model was proposed and it can handle a non-static background for a long period of training [7]. Some other techniques were proposed based on this CB model. A Box-based CB model [8] was introduced to simplify the base CB algorithm. Meanwhile [9] stated that the frequency parameter should be included into the algorithm. An improved CB model was introduced by accessing the image by clusters and not pixel by pixel [10].

In this paper we have tried to focus on the modified version and implement this CB model of background subtraction for robotic soccer vision. Color segmentation is widely used to separate the object from the soccer field. The simplest method is classifying it by using the YUV color channel [11], [12]. Classifying the color object using the YUV image also was proposed in [13] but it subtracts the background from the foreground first. This method only classified the intensity of the current image and defined the threshold value. The matching pixels will belong to background and others will belong to foreground. In this paper we have tried to implement the modified CB based background subtraction in robotic soccer vision. The aim is to acquire an inexpensive approach for object recognition for robotic soccer vision.

In the following section, the CB model and the modified CB are described. In section 3, the experimental results of the proposed approach are shown. The conclusion is addressed in section 4.

2   Codebook Based Background Subtraction

We had decided to focus on the CB method [7]. We chose this method because it is considered as a good one for background subtraction for real-time tracking. We have left out the MOG because the CB has overcome the MOG as indicated in several researches such as in [7]-[10].

2.1   Codebook

CB is a method that models the background. It is based on long scene observation to construct the model. For each pixel, it builds a codebook which consists of one or several codewords. The number of codewords for each pixel varies, depending on the background variations.

Let X be the sample value of pixel in training of sequence image: $X = \{x_1, x_2, x_3, \dots, x_N\}$ where N is the number of frames in the training sequences. Then let C as the pixel codebook consist of L codewords: $C = \{c_1, c_2, c_3, \dots, C_L\}$. Each pixel has different sizes of codebooks; it depends on the variations of the sample.

Each codeword $c_i, i = 1 \dots L$, consists of an RGB vector $v_i = (R_i, G_i, B_i)$ and 6-tuple $aux_i = \langle \hat{I}, \check{I}, f, \lambda, p, q \rangle$. Thus, the codebook model contains color, intensity and temporal information. The detailed variable of codeword is described below.

$\hat{I}$:   The maximum intensity of accepted codeword.
$\check{I}$:   The minimum intensity of accepted codeword.

f : The frequency of occurred codeword.

λ : The *maximum negative run-length* (MNRL) is defined as the longest time interval that codeword has not recurred.

p : The first access time that codeword has occurred.

q : The last access time that codeword has occurred.

In the training period, each value of x_t at time t, is compared with the current codebook to determine which the codeword c_m that matches. In CB, to determine the codeword is matching, color distortion and intensity bounds are used. The details are given below.

Algorithm 1. Codebook Modeling

$L = 0, C = \emptyset$ (empty set of codebook)

$\boldsymbol{for}\ t = 1\ to\ N\ \boldsymbol{do}$

    (i) $x_t = (R, G, B), I = \sqrt{R^2 + G^2 + B^2}$

    (ii) Find the codeword c_m in $C = \{c_i = 1 \le i \le L\}$ matching to x_i based these (a) and (b) conditions

        (a) $colordist(x_i, v_m) \le \varepsilon_1$

        (b) $intensity\big(I, \langle \hat{I}, \check{I} \rangle\big) = true$

    (iii) *If* $C = \emptyset$ or there is no match, then $L = L + 1$. Create new codeword c_L by defining:

        • $v_L = (R, G, B)$

        • $aux_L = \langle \hat{I}, \check{I}, 1, t - 1, t, t \rangle$

    (iv) Otherwise, update the matched codeword c_m consisting of $v_m = (R_m, G_m, B_m)$ and $aux_m = \langle \hat{I}_m, \check{I}_m, f_m, \lambda_m, p_m, q_m \rangle$ by defining:

        • $v_m = \left(\frac{f_m R_m + R}{f_m + 1}, \frac{f_m G_m + G}{f_m + 1}, \frac{f_m B_m + B}{f_m + 1}\right)$

        • $aux_m = \langle \max\{I, \hat{I}_m\}, \min\{I, \check{I}_m\}, f_m + 1, \max\{\lambda_m, t - q_m\}, p_m, t\rangle$

$\boldsymbol{end\ for}$

For each codeword $c_i, i = 1 \ldots L$, wrap around λ_i by defining $\lambda_i = \max\{\lambda_i (N - q_i + p_i - 1)\}$.

The condition (a) and (b) are matching when the color of x_t and c_m are close enough and the intensity of x_t lies between the range bounds of c_m. Color distortion will be described in the following. When an input pixel $x_t = (R, G, B)$ and a codeword c_i where $v_i = (R_i, G_i, B_i)$,

$||x_t||^2 = R^2 + G^2 + B^2$,

$||v_t||^2 = R_i^2 + G_i^2 + B_i^2$,

$\langle x_t, v_t \rangle^2 = (R_i R + G_i G + B_i B)^2$,

$p^2 = ||x_t||^2 \cos^2\theta = \frac{\langle x_t, v_t \rangle^2}{||v_t||^2}$

The color distortion can be calculated by

$$colordist(x_t, v_i) = \sqrt{||x_t||^2 - p^2} \qquad (1)$$

$$intensity(I, \langle \hat{I}, \breve{I} \rangle) = \begin{cases} true \ if \ I_{low} \leq ||x_t|| \leq I_{hi} \\ false \ otherwihise \end{cases} \qquad (2)$$

After the codebook model is constructed, the foreground is detected by using background subtraction directly. The detail algorithm is following:

Algorithm 2. Codebook Background Subtraction

---

1) $x_t = (R, G, B)$
2) For all codewords in M, find the codeword c_m matching to x based on these condition:
   (i) $colordist(x, c_m) \leq \varepsilon_2$
   (ii) $intensity(I, \langle \breve{I}_m, \breve{I}_m \rangle) = true$
   Update the matched codeword in Step 2 (iv) in codebook construction algorithm.
3) The pixel classified as background:

$$BGS(x) = \begin{cases} foreground \ if \ there \ is \ no \ match \\ background \ otherwise \end{cases}$$

ε_2 is detection threshold. The pixel is classified as foreground if there is no matched codeword. Otherwise it is detected as background.

2.2 Our Modified Codebook

The CB is a pixel by pixel basis algorithm. It stores all information for each pixel in codebook models, and each codebook model consists of several codewords. The codewords themselves consist of RGB vectors and 6-tuple of auxiliary parameters. Even though this CB technique has been proved to be able to overcome the performance of other methods [7], we saw that the algorithm was still complex. In robotic soccer vision, we need very fast algorithms to process images, especially in the pre-processing phase. Here, the originality of our works is to modify the CB model to be less expensive in terms of processing time. We tried to implement CB as the image segmentation part which is used for subtracting the foreground, robots and ball, from the background, which is the soccer field.

CB used the RGB image which consists of 3 channels R, G, and B. Then to get the intensity for intensity bounds, it calculated the three channels of RGB image. The calculation of intensity and color distortion from the CB algorithm is still expensive to be implemented in robotic soccer vision, because it is calculated for each pixel through the image sequences. Here, we simplify the algorithm by using images with the YUV color model. The aim is to minimize the pixel by pixel calculation. The Y image is used to get the intensity for intensity bounds. U and V are used to limit the color values. We just simply left out the color distortion and used the U and V values.

So our codeword consists of two vectors $vmin_i = (Y_{min}, U_{min}, V_{min})$ and $vmax_i = (Y_{max}, U_{max}, V_{max})$ and contains 4-tuple $aux_i = \langle f, \lambda, p, q \rangle$. Compared to the CB algorithm, our algorithm reduces the intensity calculation. Our algorithm also reduces the conditional color distortion, for which the color distortion calculation is complex enough. The details of our codebook construction algorithm are below:

Algorithm 3. Our Modified Codebook

$L = 0, C = \emptyset$ (empty set of codebook)
for $t = 1\ to\ N$ **do**
    (i) $x_t = (Y, U, V)$
    (ii) Find the codeword c_m in $C = \{c_i = 1 \leq i \leq L\}$ matching to x_i based
         these conditions
         (a) $intensity(Y, \langle Y_{min}, Y_{max}\rangle) = true$
         (b) $chromatic1(U, \langle U_{min}, U_{max}\rangle) = true$
         (c) $chromatic2(V, \langle V_{min}, V_{max}\rangle) = true$
    (iii) *If* $C = \emptyset$ or there is no match, then $L = L + 1$. Create new codeword c_L
         by defining:
         • $vmin_L = vmax_L = (Y, U, V)$
         • $aux_L = \langle 1, t - 1, t, t\rangle$
    (iv) Otherwise, update the matched codeword c_m consisting of $vmin_m = (Y_{min}, U_{min}, V_{min})$, $vmax_m = (Y_{max}, U_{max}, V_{max})$ and $aux_m = \langle \lambda_m, p_m, q_m \rangle$ by defining:
         • $vmin_m = (min(Y, Y_{min}), min(U, U_{min}), min(V, V_{min}))$
         • $vmax_m = (max(Y, Y_{max}), max(U, U_{max}), max(V, V_{max}))$
         • $aux_m = \langle f_m + 1,\ max\{\lambda_m, t - q_m\}, p_m, t\rangle$
end for
For each codeword $c_i, i = 1 \dots L$, wrap around λ_i by defining $\lambda_i = max\ \{\lambda_i\ (N - q_i + p_i - 1)\}$.

After the codebook model is constructed, the foreground is detected by using background subtraction directly. The detail algorithm is following:

Algorithm 4. Our Modified Background Subtraction

---

1) $x_t = (Y, U, V)$
2) For all codewords in M, find the codeword c_m matching to x based on these condition:
     (a) $(intensity(Y, \langle Y_{min}, Y_{max}\rangle) = true$
     (b) $chromatic1(U, \langle U_{min}, U_{max}\rangle) = true$
     (c) $chromatic2(V, \langle V_{min}, V_{max}\rangle) = true$
     Update the matched codeword in Step 2 (iv) in codebook construction algorithm.
3) The pixel classified as background:

$$BGS(x) = \begin{cases} foreground\ if\ there\ is\ no\ match \\ background\ otherwise \end{cases}$$

2.3 The YUV Color Model

Our algorithm requires YUV color images. We used the YUV color model because it separates the intensity value in the Y channel and the chromatic values in the U and V channels. Because our input image from the video and camera is in the RGB color model, it requires conversion from the RGB to the YUV image. Obtaining the YUV from the RGB image is as follows:

$$Y = 0.299 \, R + 0.587 \, G + 0.114 \, B \tag{3}$$

$$U = 0.493 * (B - Y) \tag{4}$$

$$V = 0.877 * (R - Y) \tag{5}$$

3 Experiment and Result

The performance of our proposed algorithm is the main output that we want to measure, considering that the foreground image will be used in robotic soccer vision later. We used a video which contains 900 frames of size 640 × 480 as the test sequence images. The video was captured from our robotic soccer application. The first 250 frames of the video comprise the static background in dim lighting conditions, for the codebook model construction. The next 650 frames are of objects entering the field with 2 different positions. We implemented this algorithm by using C++ and the OpenCV library. We ran the implementation on 1.83GHz Intel Pentium Core 2 Duo notebook with 3GB of RAM.

Here we have evaluated the performance of these algorithms, considering that the robotic soccer application needs really fast pre-processing image segmentation. Table 1 describes the average runtime for the proposed method. Fig. 1 is the original image to be modeled by CB methods. The resulting image after applied the methods can be referred in Fig. 2. The original CB method resulted more noise compare than our proposed method. We also tried the RGB color model with our proposed method; it also gave better performance compared to the CB model, but was worse than with our proposed method using the YUV color model.

Table 1. Performance comparison based on average runtime execution in frame/ms

Algorithms	Background Modeling	Foreground Subtraction
CB ($\varepsilon_1 = \varepsilon_2 = 1000$)	426.69	441.298
CB ($\varepsilon_1 = \varepsilon_2 = 500$)	391.562	382.905
Proposed (RGB)	27.584	11.324
Proposed (YUV)	27.161	10.983

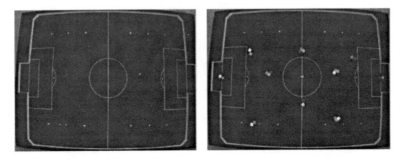

Fig. 1. Original image, the left one is with no objects and the right one is with objects

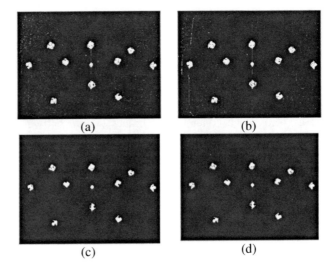

Fig. 2. Foreground detection comparison. (a) using CB with $\varepsilon_1 = \varepsilon_2 = 500$, (b) using CB with $\varepsilon_1 = \varepsilon_2 = 1000$, (c) using proposed algorithm with RGB color model, and (d) was using proposed algorithm with YUV color model.

4 Conclusion

The proposed method is a simplified CB method. The aim is to attain fast performance of image segmentation to be implanted into robotic soccer vision. The proposed method can yield good performance, and is good enough to subtract the objects in the robotic soccer field in dim lighting conditions. This proposed method also can be used for other color models such as RGB. Our future work is to implement this proposed method in our color-based object recognition for robotic soccer vision. This method will be used in the image segmentation portion to subtract the foreground (the robots and the ball) from the background (the soccer field).

References

1. Da-shan, G., Jie, Z., Le-ping, X.: A novel algorithm of adaptive background estimation. In: Proceedings of International Conference on Image Processing (2001)
2. Gordon, G., Darrell, T., Harville, M., Woodfill, J.: Background estimation and removal based on range and color. In: IEEE Computer Society Conference on Computer Vision and Pattern Recognition, vol. 2, p. 464 (1999)
3. Wren, C., Azarbayejani, A., Darrell, T., Pentland, A.: Pfinder: real-time tracking of the human body. In: Proceedings of the Second International Conference on Automatic Face and Gesture Recognition (1996)
4. Stauffer, C., Grimson, W.E.L.: Adaptive background mixture models for real-time tracking. In: IEEE Computer Society Conference on Computer Vision and Pattern Recognition, vol. 2, p. 252 (1999)
5. Dar-Shyang, L., Hull, J.J., Erol, B.: A Bayesian framework for Gaussian mixture background modeling. In: Proceedings of 2003 International Conference on Image Processing, ICIP, vol. 2, pp. III-973-6 (2003)
6. Toyama, K., Krumm, J., Brumitt, B., Meyers, B.: Wallflower: principles and practice of background maintenance. In: The Proceedings of the Seventh IEEE International Conference on Computer Vision, vol. 1, pp. 255–261 (1999)
7. Kim, K., Chalidabhongse, T.H., Harwood, D., Davis, L.: Real-time foreground-background segmentation using codebook model. Journal of Real-Time Imaging 11, 172–185 (2005)
8. Qiu, T., Yiping, X., Manli, Z.: Box-based codebook model for real-time objects detection. In: 7th World Congress on Intelligent Control and Automation, WCICA, pp. 7621–7625 (2008)
9. Ilyas, A., Scuturici, M., Miguet, S.: Real Time Foreground-Background Segmentation Using a Modified Codebook Model. In: Sixth IEEE International Conference on Advanced Video and Signal Based Surveillance, AVSS, pp. 454–459 (2009)
10. Qi, L., Chunfu, S., Hao, Y., Juan, L.: Real-time foreground-background segmentation based on improved codebook model. In: 2010 3rd International Congress on Image and Signal Processing (CISP), vol. 1, pp. 269–273 (2010)
11. Bruce, J., Balch, T., Veloso, M.: Fast and inexpensive color image segmentation for interactive robots. In: 2000 IEEE/RSJ International Conference on Intelligent Robots and Systems, vol. 3, pp. 2061–2066 (2000)
12. Lovell, N.: Machine Vision as the Primary Sensory Input for Mobile Autonomous Robot. Griffith University, Autralia (2006)
13. Wei, W., Yun-hui, L.: An illumination adaptive color objects recognition method in robot soccer match. In: IEEE International Conference on Robotics and Biomimetics, ROBIO, pp. 1200–1205 (2008)

Threaded C and Freezer OS

Jacky Baltes, Chris Iverach-Brereton, Chi Tai Cheng, and John Anderson

Autonomous Agents Lab,
University of Manitoba,
Winnipeg, Manitoba,
Canada, R3T 2N2
jacky@cs.umanitoba.ca
http://www.cs.umanitoba.ca/~jacky

Abstract. Threaded C is a meta-language that is based on C, but is annotated with thread, monitor thread, and semaphore markup. Threaded C uses the runtime provided by the Freezer OS, a small, memory-efficient embedded kernel. The combination of Freezer OS and Threaded C allows the simple expression of common control problems in robotics. The system is geared especially towards robotics education, as it matches the mental map that children have of how control structures should work.

Keywords: Scheduling, Real-time OS, Embedded Systems.

1 Introduction

This paper introduces Threaded C (TC), a meta-language that supports threads, monitors, and semaphores; and the Freezer OS, an efficient embedded kernel for the AVR AtMega128 series of micro-controllers [3].

The main contribution of TC is support for asynchronous support structures, which greatly simplifies the development of asynchronous control programs for robotics.

In this paper, we motivate the design of TC and Freezer OS by looking at the use of our system in the setting of an introductory robotics course. We demonstrate that asynchronous control structures are understood by children.

2 Motivation

Firstly, we will give an introduction into some of the unique and non-standard features of introductory programming for robots, especially for children. The authors have a lot of experience with introductory workshops for children in robotics and it is this experience that guided our design of the Freezer OS and Threaded C.

Drive Problem: The first problem we ask the students to solve is to drive the robot forward for one meter. We introduce the `DriveMotor(int s, int speed)` routines and tell them that the motors can be stopped by setting their speed to 0. The children quickly realize that this does not allow them to control the

T.-H.S. Li et al. (Eds.): FIRA 2011, CCIS 212, pp. 170–177, 2011.

distance that the robot drives. So we also introduce the `Wait(int timeout)` routine. Children then learn how increasing/decreasing the time out will drive the robot a longer/shorter distance.

The program that the students write looks as follows:

```
=== drive.c ===
void main() {
    DriveMotor( LEFT_MOTOR, 20);
    DriveMotor( RIGHT_MOTOR, 20);
    Wait( 100 );
    DriveMotor( LEFT_MOTOR, 0);
    DriveMotor( RIGHT_MOTOR, 0);
}
```

The children learn that the robot will execute sequences of instructions and learn that a simple robot has to convert timing values into distances.

Robot Dance Problem: In the second program, we ask students to develop a simple dance for the robot, which consists of straight line segments and turns. The children realize that the code is just a long string of repeating commands. We introduce subroutines to the children and tell them that by using a subroutine they can save themselves a lot of typing. We also introduce variables and tell them that they are like place holders for numbers. A typical program created by the children at this point will look as follows:

```
=== robotdance.c ===
void DriveStraight( int distance ) {
    DriveMotor( LEFT_MOTOR, 20);
    DriveMotor( RIGHT_MOTOR, 20);
    Wait( distance );
    DriveMotor( LEFT_MOTOR, 0);
    DriveMotor( RIGHT_MOTOR, 0);
}

void TurnLeft( int angle ) { ...

void TurnRight( int angle ) { ...

void main() {
  DriveStraight(30);
  TurnLeft( 120 );
  ...
}
```

Squares Problem: The next task for the children is to drive larger and larger squares using their robot and to stop when the size of the square is larger than a limit. This introduces variables, simple arithmetic operators, and `while` loops to the children.

Below we show the changes to the previous program to implement the squares subroutine:

```
=== squares.c ===
void Square( int limit ) {
   int distance = 10;
   while( distance < limit ) {
      DriveStraight( distance );
      TurnLeft( 400 );  // Results in a 90 degree turn
      distance = distance + 10
   }
}
```

Search Line Problem: The last task for the children is to make a simple line-tracking robot. The robot begins on a white surface with a path marked by a dark line. The robot must locate the line and follow it to an objective. We use this problem to teach children the three fundamental components of programming in imperative programming languages: sequence, selection, and iteration.

Fig. 1. Line Trackers are a popular entry problem for robotics and can act as navigation and localization system for more complex tasks

The first step is to simply have the robot drive an increasing square and stop when part of the robot touches the line. We introduce the int GetLightSensor() function. If the robot has an output device we also introduce a PrintNumber(int number) function to the children at this point.

This is the *crucial* part in the exercise. Many children try the following program which unfortunately does not work, since the light sensor will only be read at the end of the straight line and turn.

```
=== search_line_ideal.tc ===
void SearchLine( ) {
   int distance = 10;
   while( LightSensor() < 100 ) /* run_while */ {
      DriveStraight( distance );
      TurnLeft( 400 );  // Results in a 90 degree turn
      distance = distance + 10
   }
   FollowLine();
}
```

Even though children can understand why their program does not work, it is not easy for them to fix it. In fact, fixing this problem requires major changes to the program. Firstly, the `Wait` calls have to be replaced with loops that check the light sensor and terminate the loop if the line is found. `DriveStraight`, which previously did exactly one task, now has to do two and has to know if we are interested in the light sensor. Secondly, a return argument needs to be added which tells the caller if `DriveStraight` returned because it covered the desired distance or because the track was found.

The correct sequential version of the line searcher will look as follows assuming `int DriveStraight(int distance, int sensor)` and corresponding turn routines:

```
=== line_search_sequential.c ===
int DriveStraight( int distance, int light ) {
   int i = 0;
   DriveMotor( LEFT_MOTOR, 20);
   DriveMotor( RIGHT_MOTOR, 20);
   while( i < distance ) {
      if ( light == 1 ) {
         if ( LightSensor() >= LIMIT ) {
            DriveMotor( LEFT_MOTOR, 0); DriveMotor( RIGHT_MOTOR, 0); return 1;
         }
      }
      i = i + 1;
   }
   DriveMotor( LEFT_MOTOR, 0); DriveMotor( RIGHT_MOTOR, 0); return 0;
}
...
int SearchLine( ) {
   int distance = 10; int result = 0;

   while( result == 0 ) {
      if ( ( result = DriveStraight( distance, 1 ) ) != 0 ) {
         if ( ( result = TurnLeft( 400, 1 ) ) != 0 ) {
            distance = distance + 10
         }
      }
   }
}
...
```

The first time that we run this workshop, this complexity turned out to be too difficult for the children. The nested conditionals, scoping rules for brackets, and function values was too hard for the children to grasp in a short weekend workshop.

2.1   Threads

After realizing that children had a lot of trouble with the sequential approach, we decided a different approach in the following year. Even though threads and parallel processing are often not taught until the 3rd year of a B.Sc. degree, we introduced them to the children right from the start.

To our great surprise, children had little trouble with threads, mainly because they did not know about processors, timer interrupts, and reader-writer problems. For the children, it was easy to grasp that a computer can do more than one thing at a time. After all, children know that the real world has many things moving concurrently and they can multi-task themselves.

Therefore, in the subsequent years, we used NQC [2], a C dialect with primitive support for threads, to solve the `SearchLine` problem by introducing threads to the children and explaining how these can be started and stopped. The program then looked as follows:

```
=== line_search_threads.c ===
...
THREAD void Square( int limit ) {
   int distance = 10;
   while( distance < limit ) {
      DriveStraight( distance );
      TurnLeft( 400 );  // Results in a 90 degree turn
      distance = distance + 10
   }
}

int SearchLine( ) {
   startTask Square( 1000 );
   startTask DetectLine();

}
...
THREAD DetectLine( ) {
   if ( LightSensor() >= 100 ) {
      stopTask Square;
      startTask FollowLine();
   }
}
```

The syntax necessary to declare, start, and stop threads was at first confusing, but the general principle of the program was easy to grasp for the children. Using threads, children as young as ten were able to implement line searching and line following robots.

It is important to note that given the basic building blocks above for driving, turning, and reacting to the sensor, the children enjoyed experimenting and combining them in novel patterns. For example, changing the square pattern to a dance when looking for the line.

3   Freezer OS

After having run workshops for several years, the first author moved to another university and focused on other research areas, in particular humanoid robots[1].

Some of the robot designs were based on the Robotis Bioloid robotics kit, which uses an AVR ATMega128 micro-controller based system to control the motion.

His team developed the Freezer OS, a small, memory-efficient embedded real-time kernel for AVR-based devices. The Freezer OS is a pre-emptive multi-tasking kernel. The scheduler is a priority-based, round-robin scheduller that supports semaphores as synchronization primitives. It also included wait queues for interrupt driven devices such as the AD converter and the serial ports.

One issue in developing a kernel for a low-memory (4 KB RAM) processor such as the AtMega128 is the allocation of the stack space. Since threads may run in any order, we must allocate sufficient stack space for each thread. This fragments the memory and the system will crash if thread 1 overruns its allocated stack space, even though other threads may still have memory available.

However, as seen in the example above, often threads are only needed to check sensor readings or set outputs for an actuator. An example is reading the light sensor in the **Search Line Problem**. We therefore developed the concept of monitor threads. These threads are periodic tasks that run very quickly and do not use a large amount of stack space. Instead of reserving separate stack space for a monitor thread, a monitor thread will use the stack space of the currently running task. This means, that the monitor thread must terminate before the currently running thread can be rescheduled. This constraint is enforced in the Freezer OS scheduler that calls the monitor threads if necessary and then blocks execution of the currently running thread until the monitor thread terminates.

4   Threaded C

Previous versions of the Freezer OS kernel were used extensively to control the motions and deal with the communication of our small humanoid robots. The students at the Autonomous Agents Lab at the University of Manitoba would use standard C to start and stop tasks for example. However, manually dealing with the resources such as threads, stack space, and priorities would lead to problems. For example, students would forget to start a thread or initialize a semaphore correctly.

We therefore considered developing a new programming language to better support the Freezer OS and application development for our humanoid robots in general.

Clearly, developing a new language including compiler, libraries, and runtime is a non-trivial task. Since all of the students were comfortable programming in the C language and avr-gcc — a high quality open-source C compiler for the AVR family of processors — existed, we decided to instead implement a meta-programming language *Threaded C* (TC).

A TC program is a C program with additional markup to declare threads, monitor threads, and semaphores. A python preprocessor reads in TC files and converts them into a legal C files, which can then be compiled by any ANSI C compiler, including avr-gcc. The TC preprocessor also keeps track of the number of threads and sets global variables (e.g., `MAX_NUM_TASKS`, the maximum number

of tasks in the system) and creates a main function which will start the threads.
An example of thread markup is shown below.

```
=== search_line.tc ===
%%% THREAD priority=medium stack_size=128 %%%
void search_line( ) {
    run_while( LightSensor() < 100 ) {
        DriveStraight( distance );
        TurnLeft( 400 );  // results in a 90 degree turn
        distance = distance + 10;
    }
    FollowLine( );
    ...
}
```

The main novelty in TC is the support for asynchronous control loops using
run_while(), which is similar to a while loop, but the test will be evaluated
asynchronously and the loop will immediately if the test condition becomes false.
Internally, this is implemented as the creation of an implicit monitor thread,
allocation of a jmp_buf structure and setjmp and longjmp instructions.

Using TC, and replacing while with run_while, the ideal **Search Line
Problem** program shown above compiles and runs as expected. The TC preprocessor will take the **run_while** and convert it into the following C code. It is
important to note that the user rarely if ever has to see the C code, since the C
code is created from the TC file shown above.

```
=== search_line.c === /* DO NOT EDIT! NOT MANIPULATED BY USER */
void SearchLine( ) {
    jmp_buf jmpBuf;
    int jmpResult;
    int monitorId;

    jmpResult = setjmp(jmpBuf);
    if (jmpResult == 0) {
        monitorId = startAsyncWhileThread( &asyncMonitorFunction, &jmpBuf);
        while( LightSensor() < 100 ) {
            DriveStraight( distance );
            TurnLeft( 400 );  // Results in a 90 degree turn
            distance = distance + 10
        }
        stopMonitorThread(monitorId);
    }
    FollowLine();
    ...
}
int asyncMonitorFunction(void) {
    return (LightSensor() <= 100);
}
... In schedule(), called by Timer 1 interrupt
```

```
if ( !scheduledMonitor -> execute() ) {
    stopMonitorThread(scheduledMonitor -> tid); /* Disable all nested
monitors as well */
    task[current_task].state = READY;

    sei();
    longjmp(*(scheduledMonitor -> jmp_buf), 1);
} else {
    asm volatile("\t reti\n"::);   /* Continue the thread normally */
}
```

Particular care must be taken when dealing with nested asynchronous control structures, since the conditions run completely asynchronously and thus may terminate in any order. For example, assuming that DriveStraight uses another run_while to check a touch sensor for collisions, then it is possible that the light sensor monitor thread fires first and execution will continue with FollowLine. In this case, all monitor threads associated with the inner run_while must be terminated as well, as the context is not valid anymore. So the scheduler startAsyncWhileThread keeps saves the level of each new monitor thread. The routine stopMonitorThread will disable the current as well as all higher level monitor threads. The system then executes a longjmp with an argument of 1, which means execution will continue at line FollowLine of the ideal TC program.

5   Conclusion

The Freezer OS and TC combination works well for our applications. However, there are several extensions that we intend to work on in the future. One issue is that since the condition in the run_while is converted into its own subroutine it has no access to the local variables of its subroutine. This could be solved by also passing a pointer to the frame of the subroutine, but we have not been able to find a portable way of doing this in C.

References

1. Baltes, J., Anderson, J.: Complex AI on Small Embedded Systems: Humanoid Robotics Using Mobile Phones (2010)
2. Baum, D., Hansen, J.: Not quite c programmers guide, version 3.1r2 (2005), http://bricxcc.sourceforge.net/nqc/
3. AVR Corporation. Atmel avr atmega128 data sheet. Configurations (2010)

Taguchi Method Optimization for PCB Lamination Process

Kuo-Yang Tu and Min-Hui Li

Institute of System Information and Control,
National Kaohsiung First University of Science and Technology,
2 Juoyue Road, Nantsu, Kaohsiung City, R.O.C.
tuky@nkfust.edu.tw

Abstract. The lamination process of Printed Circuit Board (PCB) is a complex engineering system. Because there are many factors and uncertainties in the process system, it's hard for the optimization of successful lamination ratio. In this paper, the Taguchi method for lamination process optimization is proposed. Based on fundamental analysis, the major factors including board temperature, lamination temperature, lamination press, and lamination time are figured out. From the scale level of the major factors, orthogonal arrays L18 are engaged for the Taguchi experiments. Based on the orthogonal arrays L18 of the Taguchi experiments, the best successful ratio is analyzed to select the level of major factors. The selected level does again Taguchi experiment to improve the successful ratio. Experiment results show that the successful ratio improves very well.

Keywords: Dry Film, Lamination Process, Taguchi Method; S/N.

1 Introduction

The lamination process of Printed Circuit Board (PCB) is a complex engineering system. Before lamination, copper must be cleaned and micro-etching to coarsen its surface, laminates dry film on it in adaptive temperature and press. In the lamination process, successful ratio is influenced by temperature and press in every step. To find adaptive temperature and press used in every step is very important but complicated.

Traditional method adjusts the process based on a proposed combination via designer experiments. However, this method doesn't analyze the process so that the designer never understands how to do for the best. Thus, in this paper, the PCB lamination process optimization based on Taguchi experiment method is proposed.

During PCB lamination process, the successful ratio is influenced by many control factors. To solve optimal combination based the control factors is very important to promote the throughput of the PCB lamination. After finding the optimization parameters, the better successful ratio is confirmed by experiment again. Higher successful ratio is better lamination process. Thus the performance index makes use of larger-to-better. In the experiments, S/N (Signal-to-Noise) is used as the influence degree of the major factors.

T.-H.S. Li et al. (Eds.): FIRA 2011, CCIS 212, pp. 178–185, 2011.
© Springer-Verlag Berlin Heidelberg 2011

2 System Description

Figure 1 is the photo of a PCB (Printing Circuit Board) lamination process equipment. Usually such equipment sends PCB to lamination machine via rollers. During sending, the lamination machine heats and press to let copper stick on the PCB. The copper must perfect to stick on the PCB. Otherwise it will be a failure PCB, and must be discarded. Successful ratio is very important of the lamination process usage. However, it's difficult to handle the factor influencing the successful ratio.

Control factors influencing PCB lamination process are hard to solve based on traditional methods. Thus genetic algorithm, neural network, Taguchi method, etc, is proposed for optimal control factors. In this paper, the optimal PCB lamination process based neural network model via Taguchi experiment method is proposed. The main advantage of Taguchi experiment method is to use few set of experiment combination for the optimization trend. Compared with traditional method that usually uses all control factors, it's an effective method for optimization.

The experiment steps of Taguchi method are:

1. Define objective function,
2. Extract the control factors from the defined objective function,
3. Scale the content and quantity of the control factors,
4. Define the content and quantity of controllable factors,
5. Design orthogonal table for complete experiments,
6. Experiments,
7. Confirm the experiments for optimization.

Fig. 1. The photo of lamination process machine

Taguchi orthogonal array design is based on the following formula:

$$L^a(b^c \times d^e) \tag{1}$$

where a is the numbers of experiment set, c is the numbers of b levels, and e is the number of d levels. The meaning of this formula is that the Taguchi orthogonal array L^a can include that the numbers of b-level and d-level are c and e, respectively.

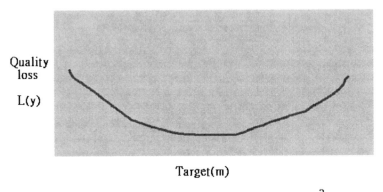

Quality loss

L(y)

Target(m)

Quality Characteristics : $L(Y)=k*(y-m)^2$

Fig. 2. Quality loss

During the optimization process, it's important to find quality characteristics of objective function. Taguchi defined quality loss to describe the payment of product quality during its life cycle.   Lower quality loss represents higher quality characteristics.   In general, the quality loss is described by second order function. Fig. 2 is a kind of quality loss function.   In this Fig, the optimal quality is the minimum loss at m.   The loss increases when the quality is far away from the m.

Usually, the quality loss is defined by an objective function as follows:

$$\text{Quality loss} = \frac{1}{n}\Sigma_{i=1}^{n}k\frac{1}{(y_i - m)^2} \tag{2}$$

$$\text{Average quality loss} = k[(\bar{y} - m)^2 + s^2] \tag{3}$$

Mean Square Deviation (MSD) is a very popular method to describe quality characteristics. There are three kinds of MSD as follows:

$$\text{Target quality } MSD_t = [(\bar{y} - m)^2 + s^2] \tag{4}$$

$$\text{Minimum quality } MSD_s = \bar{y} + s^2 \tag{5}$$

$$\text{Maximum quality } MSD_b = \frac{1}{n}\Sigma_{i=1}^{n}\frac{1}{y_i^2} \tag{6}$$

Where the optimal MSD_s is a minimum value, but the optimal MSD_b is a maximum value the target quality of MSD_t is m.

In addition, Signal-to-Noise is the evaluation standard of robustness signal. It becomes very useful output of Taguchi experiment result, and is engaged to select the optimization parameters in orthogonal arrays. In a word, the Taguchi method executes experiments, collects data and conduct data analysis. The analysis results are shown by S/N ratios in response tables and response graphics. Finally, confirmation run make sure the selected optimization parameters.

3 Taguchi Experiments

The Taguchi experiment method adapts L^{18} orthogonal array including four important factors, board temperature, lamination temperature, lamination press, and lamination time. Two different board temperature (20 °C and 40 °C), three different lamination temperature (50 °C, 60 °C and 70 °C), three different lamination press (3 kg/cm^2, 4 kg/cm^2 and 5kg/cm^2) and three different lamination time (5 S, 6 S and 7 S) are combined in the experiments. In total, the numbers of experiments are 54 ($2^1*3^3 =$ 54) that is too huge to conduct for data analysis. Thus Taguchi experiment method provides orthogonal arrays to reduce the experiment number.

Table 1. Orthogonal array L^{18} ($2^1 \times 3^7$)

factor / Time	A	B	C	D
1	A_1	B_1	C_1	D_1
2	A_1	B_1	C_2	D_2
3	A_1	B_1	C_3	D_3
4	A_1	B_2	C_1	D_1
5	A_1	B_2	C_2	D_2
6	A_1	B_2	C_3	D_3
7	A_1	B_3	C_1	D_2
8	A_1	B_3	C_2	D_3
9	A_1	B_3	C_3	D_1
10	A_2	B_1	C_1	D_3
11	A_2	B_1	C_2	D_1
12	A_2	B_1	C_3	D_2
13	A_2	B_2	C_1	D_2
14	A_2	B_2	C_2	D_3
15	A_2	B_2	C_3	D_1
16	A_2	B_3	C_1	D_3
17	A_2	B_3	C_2	D_1
18	A_2	B_3	C_3	D_2

Table 2. The experimental result of L^{18} orthogonal array

Time	A	B	C	D	y	MSD	H
1	20	50	3	5	6.666667	0.022865	16.40834
2	20	50	4	6	7	0.020408	16.90196
3	20	50	5	7	7.333333	0.018814	17.25524
4	20	60	3	5	6.666667	0.022865	16.40834
5	20	60	4	6	7.666667	0.017219	17.63982
6	20	60	5	7	7.333333	0.018814	17.25524
7	20	70	3	6	6.333333	0.025321	15.96515
8	20	70	4	7	7	0.02127	16.72226
9	20	70	5	5	6.666667	0.022865	16.40834
10	40	50	3	7	7.333333	0.018814	17.25524
11	40	50	4	5	7.666667	0.017219	17.63982
12	40	50	5	6	7.666667	0.017219	17.63982
13	40	60	3	6	7.333333	0.018814	17.25524
14	40	60	4	7	7.666667	0.017219	17.63982
15	40	60	5	5	8.333333	0.014532	18.37678
16	40	70	3	7	7.666667	0.017219	17.63982
17	40	70	4	5	8	0.016126	17.92466
18	40	70	5	6	7.666667	0.017219	17.63982

In the Taguchi experiment method, let board temperature be A = {A1, A2} = { 20°C, 40°C}, lamination temperature be B = {B$_1$, B$_2$, B$_3$} = {50 °C, 60°, 70 °C}, Lamination press be C = {C$_1$, C$_2$, C$_3$} = {3 kg/cm^2, 4 kg/cm^2, 5kg/cm^2} and lamination time D = {D$_1$, D$_2$, D$_3$} = {5S, 6S, 7S}. From Eq. (1), orthogonal array L^{18} can consist of one two-different degree and seven three-different degrees. Under Taguchi method, the data analysis engages 18 experiments to approach 54 experiments.

The objective of Taguchi experiments is to find maximum successful ratio. Higher successful ratio owns better the performance. Thus the experiments engage Larger-the-Better. The experiments extract successful factors for higher successful ratio and higher signal-to-noise.

Table 1 is the experiment structure of L^{18} orthogonal array. A, B, C, and D are board temperature, lamination temperature, lamination press and lamination time, respectively. Every experimental result is written in the table.

Table 2 is the experiment result of L^{18} orthogonal array. Let y be the average successful times in 30 experiments. Then, in the table, MSD and η are shown in last two columns, respectively, where

$$MSD = \frac{1}{n} \sum_{i=0}^{n} \frac{1}{y_i^2} \text{, and}$$

$$\eta = -10 * \log(MSD). \tag{7}$$

According to the data analysis, Table 2 can obtain response table as shown in Table 3. In the response table, raw is major factors including board temperature, lamination temperature, lamination press, and lamination time. The column is major factors 1, 2 and 3. They are according to the maximum values of A_2, B_2, C_3 and D_3. Thus, the major factors that can obtain maximum η are board temperature 40 °C, lamination temperature 60 °C, lamination press 5kg/cm^2 and lamination time 7 seconds. However, they must be confirmed by other experiments.

Table 3. Response Table

Major Factors	Board Temperature	Lamination Temperature	Lamination Press	Lamination Time
1	16.77386	13.9101	16.82202	17.19438
2	17.66789	17.35534	17.41139	17.17364
3		17.05001	17.42921	17.2946
Distance	0.894036	3.445241	0.607184	0.120968

From the observation of Table 3, the best η is 17.66789 at $A_2 = 40$ °C, is 17.35534 at $B_2 = 60$ °C, is 17.42921 at $C_3 = 5$kg/cm2, and is 17.2946 at $D_3 = 7$ seconds. Therefore, the optimization parameters of major factors are on $A_2 = 40$ °C, $B_2 = 60$ °C, $C_3 = 5$kg/cm2, and $D_3 = 7$ seconds, respectively. Therefore, according to the optimization parameters the confirmation run executes experiment again. And the result is shown in Table 4. As shown in table 4, the average of successful ratio is 73.3% in origin parameters, but boosts to 83.3% in optimal parameters. In addition, the ratio of S/N, 18.37678, as shown in Table 2 represents that this experiment has high reappearing probability.

In original, the experiment parameters are board temperature 20 °C, lamination temperature 50 °C, lamination press 4 kg/cm^2 and lamination time 6 seconds. In the experiment result, the average of successful ratio is 73.3%, and S/N is 17.25524. In this paper, the Taguchi experiment method selects optimal parameters, board temperature 40 °C, lamination temperature 60 °C, lamination press 5 kg/cm^2 and lamination time 7 seconds. The optimal parameters result in 83.3% average successful ratio, and 18.37678 S/N. The successful ratio and S/N increase 10% and

Table 4. The experiment result of confirmation run

Factor	Origin		Optimization	
	scale	Success	Scales	Success
A	20℃	7	40℃	8
B	50	8	60℃	9
C	4kg/cm^2	7	5kg/ cm^2	8
D	6s		7s	

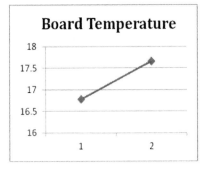

Fig. 3. Response graphic during board temperature change

1.121538, respectively. Such results show that it's very useful and highly efficiency to only run 18 experiments in Taguchi method for the study of needing 54 experiments in total.

Response graphic is also a very useful to observe quality characteristics. For example, Fig. 3 is the response graphic of board temperature. This response graphic being high slope has bad characteristics. As a result, the optimal board temperature changes to 40 °C.

4   Conclusions and Further Development

In this paper, Taguchi experiment method for optimal parameters of PCB lamination process is proposed. The Taguchi method focuses on searching the optimal parameters of four causal factors, board temperature, lamination temperature, lamination press, and lamination time. Compared with the parameters based on designer experience, the optimal parameters searched by Taguchi, method increases 10% successful ratio, and 1.12153 S/N. This experiment result demonstrates the excellent performance of Taguchi method for complicated PCB lamination process. Especially, the total experiments need 54 times in the lamination process, but the Taguchi method only uses 18 time experiments. Such result shows that the Taguchi method to search optimal parameters is very effective. It can be used to reduce the cost of finding optimal process parameters. However, the scales between two levels are too big such as board temperature has two levels, 20 °C and 40 °C. It is possible to imply other variation between these two levels. The other method to find the optimal parameters between rough level is the further development.

Acknowledgments. This research was partly supported by National Science Council, Taiwan, Rep. of China under grant NSC 97-2221-E-327-019-.

References

[1] Vasiljevic, Darko: Optimization of the Cooke triplet with the various evolution strategies and the damped least squares. In: Proceedings of SPIE – The International Society for Optical Engineering, vol. 3780, pp. 207–215 (1999)

[2] Tsai, J.T., Liu, T.K., Chou, J.H.: Hybrid Taguchi-genetic algorithm for global numerical optimization. IEEE Transactions on Evolutionary Computation 8(4), 365–377 (2004)

[3] Chou, J.H., Liao, W.H., Li, J.J.: Application of Taguchi-Genetic Method to Design Optimal Grey-Fuzzy Controller of a Constant Turning Force System. In: Proc. of the 15th CSME Annual Conference, Taiwan, pp. 31–38 (1998)

[4] Goldberg, D.: Genetic Algorithms in Search. In: Optimization and Machine Learning, pp. 151–159. Addison-Wesley, Reading (1989)

[5] Wu, Y.: Taguchi Methods for Robust Design. The American Society of Mechanical Engineers, New York (2000)

[6] Leo Zhang, K., Wei, D., Catabay, W., Wang, Y., Dong, J., Yoshikawa, S., Ho, Y.L.: A Correlation of PECVD Oxide Film Charge to Transistor Leakage. In: Proceedings of VMIC Conference, p. 275 (2000)

[7] Taguchi, G.: (Yuin Wu, technical editor for the English edition), Taguchi Methods / Design of Experiments, Dearborn. MI / ASI Press, Tokyo

[8] Hedayat, A.S., Sloane, N.J.A., Stufken, J.: Orthogonal Arrays: Theory and Applications. Springer, New York (1999)

Parameter Estimation of Potential Field Method with Fuzzy Control for Motion Planning of Soccer Robot

Li-Chun Lai[1], Chun-Feng Lu[1], Yen-Ching Chang[2], and Tsong-Li Lee[3]

[1] Department of Electrical Engineering, Chung Chou Institute of Technology,
Yuanlin, Changhua 510, Taiwan, R.O.C.
{lclai,cflu}@dragon.ccut.edu.tw
[2] Department of Applied Information Sciences, Chung Shan Medical University,
Taichung 402, Taiwan, R.O.C.
nicholas@csmu.edu.tw
[3] Department of Automation Engineering, Nan Kai University of Technology,
Nantou 542, Taiwan, R.O.C.
alee88@nkut.edu.tw

Abstract. Based on fuzzy control to adjust attractive factor and repulsive factor of potential function, a method is proposed to navigate a soccer robot from a given initial configuration to a ball and avoid an opponent robot. The potential field method has some problems that include oscillations in the presence of obstacles. Thus, this paper suggests use fuzzy control to adjust attractive factor and repulsive factor. By the attractive factor and repulsive factor of the potential function, the navigation path of the mobile robot is smooth and the shortest. To show the feasibility of the proposed method, the simulation results are included in the following illustrations.

Keywords: Fuzzy Control, Soccer Robot, Potential Field, Motion Planning.

1 Introduction

Since the complexity of robots' performing tasks grows rapidly, research and development of multi-robot system are strongly needed. Application of multiple mobile robots exhibiting cooperative behaviors can be found in tasks such as factory automation, cleaning hazardous wastes, and moving furniture. From the standpoint of multi-robot systems, soccer robots can be considered as a common test-bench and research subjects will include collision [1], motion planning [2], control architecture [3,4], motion control [5], role selection [6], action selection [7,8], vision tracking system [9], communication [10] and so on.

Motion planning is an important issue in mobile robotics. In an environment with obstacles, the aim of path planning is to find a suitable collision-free path for a mobile robot to move from a given initial configuration to a desired final configuration.

The earliest algorithms for motion planning problems of robots deal with navigation of a robot in a completely known environment filled with stationary obstacles [11-14]. Another kind of motion planning algorithms deals with navigation of a robot in a completely unknown environment [15-17]. Since the environment is unknown to the mobile robot, different sensors such as computer vision, ultrasonics,

T.-H.S. Li et al. (Eds.): FIRA 2011, CCIS 212, pp. 186–192, 2011.
© Springer-Verlag Berlin Heidelberg 2011

and odometers will be used in these algorithms, and each of these algorithms shows its feasibility in different application areas.

The basic concept is to fill the working space of the robot with an artificial potential field in which the robot is attracted to its goal position and is repulsed away from the obstacles [18]. This potential field method is particularly attractive because of its mathematical elegance and simplicity, and has been studied extensively for mobile robot motion planning in the past decade [12-14, 18-19]. However, this method has some problems such as oscillations in the presence of obstacles. In this paper, the fuzzy control is proposed to determine attractive factor and repulsive factor of the potential field function; then, the path of the mobile robot is smooth from a given initial configuration to the ball.

The remaining sections of this paper are organized as follows: Section 2 shows the potential field method. Section 3 presents a fuzzy-based potential field method for soccer robot navigation. Simulations are performed in Section 4 to confirm the feasibility of the proposed algorithm. Section 5 concludes the paper.

2 Potential Field Method for Motion Planning

In the previous research of potential field methods, for simplicity, a robot is usually taken as a point. If the position of the soccer robot is denoted by $[x_r, y_r]$, then the most commonly used attractive potential field between a robot and an obstacle takes the following form Latombe [18]

$$U_{att} = \frac{1}{2}\xi[(x_r - x_b)^2 + (y_r - y_b)^2] \tag{1}$$

where ξ is a positive scaling factor and $[x_b, y_b]$ denotes the position of the ball point. The corresponding attractive force is then given by the negative gradient of the attractive potential field.

$$\mathbf{F}_{att} = -\nabla U_{att} = \xi \begin{bmatrix} x_b - x_r \\ y_b - y_r \end{bmatrix} \tag{2}$$

which converges linearly toward zero as the soccer robot approaches the ball. On the other hand, one commonly used repulsive potential field between a soccer robot and an opponent robot takes the following form Latombe:

$$U_{rep} = \begin{cases} \frac{1}{2}\eta(\dfrac{1}{\sqrt{(x_r - x_o)^2 + (y_r - y_o)^2}} - \dfrac{1}{\rho})^2, & \text{if } \sqrt{(x_r - x_o)^2 + (y_r - y_o)^2} \le \rho \\ 0, & \text{otherwise} \end{cases} \tag{3}$$

where η is a positive scaling factor, ρ is a positive constant denoting the distance of influence of the opponent robot, and $[x_o, y_o]$ denotes the point on the opponent robot such the distance $\sqrt{(x_r - x_o)^2 + (y_r - y_o)^2}$ is minimal. The corresponding repulsive force is then given as follows:

$$\mathbf{F}_{rep} = -\nabla U_{rep} = \begin{cases} \begin{bmatrix} F_{rep,x} \\ F_{rep,y} \end{bmatrix}, & \text{if } \sqrt{(x_r - x_o)^2 + (y_r - y_o)^2} \leq \rho \\ 0, & \text{otherwise} \end{cases} \quad (4)$$

$$F_{rep,x} = \eta(\frac{1}{\sqrt{(x_r - x_o)^2 + (y_r - y_o)^2}} - \frac{1}{\rho}) \cdot \frac{(x_r - x_o)}{\sqrt{(x_r - x_o)^2 + (y_r - y_o)^2}} \\ \cdot \frac{1}{(x_r - x_o)^2 + (y_r - y_o)^2} \quad (5)$$

$$F_{rep,y} = \eta(\frac{1}{\sqrt{(x_r - x_o)^2 + (y_r - y_o)^2}} - \frac{1}{\rho}) \cdot \frac{(y_r - y_o)}{\sqrt{(x_r - x_o)^2 + (y_r - y_o)^2}} \\ \cdot \frac{1}{(x_r - x_o)^2 + (y_r - y_o)^2} \quad (6)$$

The total force applied to the robot is the sum of the attractive force and the repulsive force

$$\mathbf{F}_{total} = \mathbf{F}_{att} + \mathbf{F}_{rep} \quad (7)$$

which will determine the movement of the soccer robot.

Overview of the potential field is depicted as shown in Figure 1.

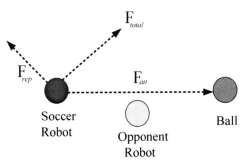

Fig. 1. Overview of the potential field method

3   Determination of the Attractive Factor and Repulsive Factor with Fuzzy Control

The attractive factor and the repulsive factor play an important role in the proposed potential field method. If the attractive factor and repulsive factor of potential field are not well, then soccer robot will show local minima and oscillations in the presence of obstacles problem. The fuzzy control determine attractive factor and repulsive factor of the potential field function, then the path of the mobile robot is smooth from a given initial configuration to the ball.

In designing the fuzzy navigation algorithm, without loss of generality, one can assume that the shape of the membership functions is triangular and symmetric. Meanwhile, it can also be assumed that there are five membership functions for each of the input linguistic variables d_{rb} and d_{ro}, respectively. d_{rb} denotes the distance of $[x_r, y_r]$ to $[x_b, y_b]$. d_{ro} denotes the distance of $[x_r, y_r]$ to $[x_o, y_o]$. The membership functions of d_{rb} will be denoted as VS (Very Small), S (Small), M (Medium), B (Big), and VB (Very Big), respectively. The membership functions of d_{ro} will be denoted as VS (Very Small), S (Small), M (Medium), B (Big), and VB (Very Big), respectively. There are five membership functions for each of the output linguistic variables ξ and η, respectively. Since there are five membership functions for d_{rb} and d_{ro}, respectively.

In determining the ξ and η of input variables d_{rb} and d_{ro}, the method of center-of-area [20] will be used.

4   Simulation Results

In this simulation example, it will be shown how to design fuzzy control to determine the attractive factor and repulsive factor when performing the proposed navigation algorithm, the value v and ρ are chosen to 10 cm/s and 6 cm, respectively. The position of the soccer robot is (100cm, 73cm). The position of the ball is (200cm, 75cm). The position of the opponent robot is (150cm, 75cm). The attractive factor and the repulse factor of the potential field method are chosen to be 10 and 2; the simulation result is shown in Figure 2. It is applying fuzzy control in Figure 5. The membership functions of d_{rb} and d_{ro} as shown in Figure 3(A) and Figure 3(B). The membership functions of ξ and η as shown in 3 (C) and Figure 3(D). The fuzzy rules in Figure 4 will be used.

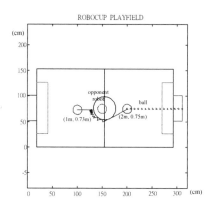

Fig. 2. The navigation distance of soccer robot is 240cm from the initial configuration (100cm, 73cm) to the final configuration (200cm, 75cm) with the attractive factor and the repulse factor of the potential field method are chosen to be 10 and 2.

(A) Membership functions of d_{rb}

(B) Membership functions of d_{ro}

(C) Membership functions of ξ

(D) Membership functions of η

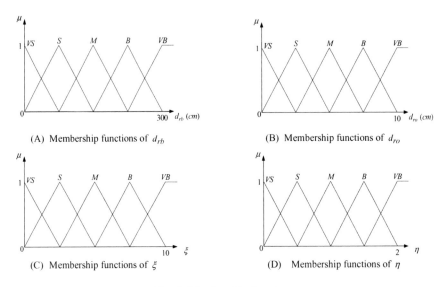

Fig. 3. Membership functions

1. If d_{rb} is *VS* and d_{ro} is *VS*, then ξ is *VS* and η is *VB*
2. If d_{rb} is *VS* and d_{ro} is *S*, then ξ is *S* and η is *B*
3. If d_{rb} is *VS* and d_{ro} is *M*, then ξ is *M* and η is *M*
4. If d_{rb} is *VS* and d_{ro} is *B*, then ξ is *B* and η is *VS*
5. If d_{rb} is *VS* and d_{ro} is *VB*, then ξ is *VB* and η is *VS*
6. If d_{rb} is *S* and d_{ro} is *VS*, then ξ is *S* and η is *VB*
7. If d_{rb} is *S* and d_{ro} is *S*, then ξ is *M* and η is *M*
8. If d_{rb} is *S* and d_{ro} is *M*, then ξ is *B* and η is *S*
9. If d_{rb} is *S* and d_{ro} is *B*, then ξ is *VB* and η is *VS*
10. If d_{rb} is *S* and d_{ro} is *VB*, then ξ is *VB* and η is *VS*
11. If d_{rb} is *M* and d_{ro} is *VS*, then ξ is *M* and η is *VB*
12. If d_{rb} is *M* and d_{ro} is *S*, then ξ is *B* and η is *M*
13. If d_{rb} is *M* and d_{ro} is *M*, then ξ is *VB* and η is *VS*
14. If d_{rb} is *M* and d_{ro} is *B*, then ξ is *VB* and η is *VS*
15. If d_{rb} is *M* and d_{ro} is *VB*, then ξ is *VB* and η is *VS*
16. If d_{rb} is *B* and d_{ro} is *VS*, then ξ is *B* and η is *VB*
17. If d_{rb} is *B* and d_{ro} is *S*, then ξ is *B* and η is *M*
18. If d_{rb} is *B* and d_{ro} is *M*, then ξ is *VB* and η is *VS*
19. If d_{rb} is *B* and d_{ro} is *B*, then ξ is *VB* and η is *VS*
20. If d_{rb} is *B* and d_{ro} is *VB*, then ξ is *VB* and η is *VS*
21. If d_{rb} is *VB* and d_{ro} is *VS*, then ξ is *VB* and η is *VB*
22. If d_{rb} is *VB* and d_{ro} is *S*, then ξ is *VB* and η is *B*
23. If d_{rb} is *VB* and d_{ro} is *M*, then ξ is *VB* and η is *VS*
24. If d_{rb} is *VB* and d_{ro} is *B*, then ξ is *VB* and η is *VS*
25. If d_{rb} is *VB* and d_{ro} is *VB*, then ξ is *VB* and η is *VS*

Fig. 4. The fuzzy rules, where the membership functions are defined as shown in Figure 3(A), (B), (C), (D), respectively.

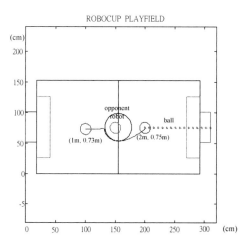

Fig. 5. The navigation distance of soccer robot is 120m from the initial configuration (100cm, 73cm) to the final configuration (200cm, 75cm) with fuzzy control

The navigation distance of soccer robot is 240cm and 120cm in Figure 2 and Figure 5, respectively. The shortest path of the soccer robot is obtained in Figure 5 with fuzzy control.

5 Conclusions

The paper uses fuzzy control to adjust the attractive factor and the repulse factor of potential function, a method is proposed to navigate a soccer robot from a given initial configuration to a ball in an unknown environment filled with opponent robot. Compared with most other potential field function, the fuzzy control can solve oscillations in the presence of obstacles. The simulation results can show the feasibility of the proposed method.

Acknowledgment. This work was supported in part by the Chung Chou Institute of Technology, Taiwan, R.O.C.

References

1. Vendittelli, M., Laumond, J.P.: Obstacle Distance for Car Like Robots. IEEE Transactions on Robots and Automation 15, 678–691 (1999)
2. Kim, J.H., Kim, K.C., Kim, D.H., Kim, Y.J., Vadakkepat, P.: Path Planning and Role Selection Mechanism for Soccer Robots. In: IEEE International Conference on Robotics and Automation, pp. 3216–3221. IEEE Xplore Press, Leuven (1998)
3. Lee, C.C.: Fuzzy Logic in Control Systems: Fuzzy Logic Controller-Part I, II. IEEE Transactions on System, Man, and Cybern. 20, 404–435 (1990)
4. Zimmermann, H.J.: Fuzzy Set Theory and Its Applications. Kluwer Academic, Boston (1991)

5. Lee, S., Bautista, J.: Motion Control for Micro-robot Playing Soccer Games. In: IEEE International Conference on Robotics and Automation, pp. 2599–2604. IEEE Xplore Press, Leuven (1998)
6. Kim, J.H., Shim, H.S., Kim, H.S., Jung, M.J., Vadakkepat, P.: Action Selection and Strategies in Robot Soccer System. In: IEEE International Conference on Circuits and Systems, pp. 518–521. IEEE Xplore Press, Sacramento (1998)
7. Kim, H.S., Shim, H.S., Jung, M.J., Kim, J.H.: Action Selection Mechanism for Soccer Robot. In: IEEE International Conference on Robotics and Automation, pp. 390–395. IEEE Xplore Press, Sacramento (1997)
8. Wu, C.J., Lee, T.L.: A Fuzzy Mechanism for Action Selection of Soccer Robots. Journal of Intelligent and Robotic Systems 39, 57–70 (2004)
9. Hong, C.S., Chun, S.M., Lee, J.S., Hong, K.S.: A Vision-guided Object Tracking and Prediction Algorithm for Soccer Robots. In: IEEE International Conference on Robotics and Automation, pp. 346–351. IEEE Xplore Press, Sacramento (1997)
10. Wang, J.: On sign-board Based Inter-robot Communication in Distributed Robotic System. In: IEEE International Conference on Robotics and Automation, pp. 1045–1050. IEEE Xplore Press, San Diego (1994)
11. Conte, G., Longhi, S., Zulli, R.: Robot Motion Planning for Unicycle and Car-like Robots. International Journal of Systems Science 27, 791–798 (1996)
12. Ge, S.S., Cui, Y.J.: New Potential Function for Mobile Robot Path Planning. IEEE Transactions on Robotics and Automation 16, 615–620 (2000)
13. Tsourveloudis, N.C., Valavanis, K.P., Hebert, T.: Autonomous Vehicle Navigation Utilizing Electrostatic Potential Fields and Fuzzy Logic. IEEE Transactions on Robotics and Automation 17, 490–497 (2001)
14. Song, P., Kumar, V.: A potential Field Based Approach to Multi-robot Manipulation. In: IEEE International Conference on Robotics and Automation, pp. 1217–1222. IEEE Xplore Press, Washington, DC (2002)
15. Lee, T.L., Lai, L.C., Wu, C.J.: A Fuzzy Algorithm for Navigation of Mobile Robots in Unknown Environments. In: IEEE International Symposium on Circuits and Systems, vol. 4, pp. 3039–3042 (2005)
16. Mester, G.: Obstacle Avoidance of Mobile Robots in Unknown Environments. In: 5th International Symposium on Intelligent Systems and Informatics, pp. 123–127. IEEE Xplore Press, Subotica (2007)
17. Toibero, J.M., Carelli, R., Kuchen, B.: Switching Control of Mobile Robots for Autonomous Navigation in Unknown Environments. In: IEEE International Conference on Robotics and Automation, pp. 1974–1979. IEEE Xplore Press, Roma (2007)
18. Latombe, J.: Robot Motion Planning. Kluwer, Boston (1991)
19. Lai, L.C., Wu, C.J., Shiue, Y.L.: A Potential Field Method for Robot Motion Planning in Unknown Environments. Journal of the Chinese Institute of Engineers 30, 369–377 (2007)
20. Lin, C.T., Lee, C.S.G.: Neural Fuzzy Systems: A Neural-Fuzzy Synergism to Intelligent Systems. Prentice-Hall, NJ (1996)

A Fast Identification Algorithm with Skewness Noises under Box-Cox Transformation-Based Annealing Robust Fuzzy Neural Networks

Pi-Yun Chen[1], Yu-Yi Fu[2], Jin-Tsong Jeng[3], and Kuo-Lan Su[4]

[1] Department of Electrical Engineering, National Chin-Yi University of Technology, Taichung, Taiwan
[2] Department of Automation Engineering, Nan-Kai University of Technology, Nantou, Taiwan
[3] Department of Computer Science and Information Engineering, National Formosa University, Yunlin, Taiwan
(tsong@nfu.edu.tw)
[4] Department of Electrical Engineering, National Yunlin University of Science and Technology, Yunlin, Taiwan

Abstract. This paper proposes Box-Cox transformation-based annealing robust fuzzy neural networks (ARFNNs) that can be used effectively for function approximated problem with skewness noises. In order to overcome the skewness noises problem, the Box-Cox transformation that its object is usually to make residuals more homogeneous in regression, or transform data to be normally distributed has been added to the annealing robust fuzzy neural networks. That is, the proposed approach uses Box-Cox transformation for skewness noises problem and support vector regression (SVR) for the number of rule in the simplified fuzzy inference systems. After the initialization, an annealing robust learning algorithm (ARLA) is then applied to adjust the parameters of the Box-Cox transformation-based annealing robust fuzzy neural networks. Simulation results show that the proposed approach has a fast convergent speed and more generalization capability for the function approximated problem with skewness noises.

Keywords: annealing robust fuzzy neural networks, Box-Cox transformation, support vector regression, skewness noises problem.

1 Introduction

Determining the model between inputs and outputs is an important theme for many engineering problems, which usually requires some desired output be presented by a given set of input variables. Unfortunately the data have exhibited combinations of skewness noises. It may occur due to various reasons, such as erroneous measurements or noisy data from the tail of noise distribution functions. The most commonly used family is the modified power transformations of Box and Cox [1]. In nonlinear problem, a model for the expectation of the response is often available from a theoretical understanding of the system giving rise to the data, through preliminary

T.-H.S. Li et al. (Eds.): FIRA 2011, CCIS 212, pp. 193–201, 2011.
© Springer-Verlag Berlin Heidelberg 2011

data analysis. However, the input-output relation of the nonlinear system model may not be well understood. Thus, the nonlinear system may be approximated by a given set of input-output training data. In recent years, several identification methods such as the neural networks [2] and the fuzzy modeling [3] for the nonlinear systems have been proposed. A fuzzy system consists of fuzzy rules and fuzzy inference, has been proved to be a university approximator [4]. It has many applications for the efficient identification and control of nonlinear systems. Therefore, there are many researches in mixes neural network and fuzzy logic to be a fuzzy neural network [5]. However, the number of rule in the simplified fuzzy inference system and the initial weights of structure for the fuzzy neural networks are difficult to determine at the same time. In addition, the obtained training data sometimes contain skewness noises. Hence, when the skewness noises come off, there are some issues in the traditional approaches [6].

Hong [7] proposed that a Box-Cox transformation-based radial basis function networks (RBFNs) model base is derived based on a rank revealing orthogonal matrix triangularization; namely, QR decomposition [8]. Besides, the identification algorithm uses Gauss-Newton method to derive the Box-Cox transformation parameter. For a large data set, using the QR decomposition increases computational expense for model structure, and when the skewness noises are exists, there still exist some problems in this algorithm. In this paper, in order to overcome the above problem, we demonstrate the Box-Cox transformation-based ARFNNs with SVR that can be used effectively for the function approximation and the identification algorithm of the nonlinear systems with skewness noises. The SVR method with Gaussian kernel function is proposed to improve the initial structure of the fuzzy neural networks. It has been proved that the SVR is applied to determine the number of rule in the simplified fuzzy inference systems and the solutions of the SVR are used as initial weights in the fuzzy neural networks at the same time, based on these initial conditions, the proposed fuzzy neural networks has been proved to have a fast convergent speed [9]. After initialization, an annealing robust learning algorithm is then applied to adjust the parameters of the Box-Cox transformation-based ARFNNs.

2 The Box-Cox Transformation-Based ARFNNs

Assume that the unknown nonlinear system with skewness noises ξ is expressed by

$$y(t+1) = f(y(t), y(t-1), \cdots, y(t-n), u(t), u(t-1), \cdots, u(t-m)) + \xi, \quad (1)$$

where $y(t)$ is the output of the system, $u(t)$ is the input of the system, $f(\cdot)$ is the unknown nonlinear function, and n and m are the structure orders of the system. The objective of Box-Cox transformation is usually to make residuals more homogeneous in regression, or transform data to be normally distributed. The well known Box-Cox version of power transformation [1] is formed as

$$z(y, \lambda) = \begin{cases} \dfrac{y^{\lambda} - 1}{\lambda \tilde{y}^{\lambda-1}}, & \text{if } \lambda \neq 0 \\[2mm] \tilde{y} \log y, & \text{if } \lambda = 0 \end{cases}, \quad (2)$$

where λ is the transformation parameter and $\tilde{y} = \sqrt[N]{\prod_{i=1}^{N} y_i}$ is the geometric mean of the system output. For a given $\lambda \neq 0$, the system output $y(t+1)$ is replaced by the Box-Cox transformation (BCT) normalized response $z(t+1)$ using (2). That is,

$$z(t+1) = BCT(y(t+1), \lambda) = \frac{y(t+1)^{\lambda} - 1}{\lambda \tilde{y}(t+1)^{\lambda-1}} \tag{3}$$

Function $\bar{f}(\cdot)$ is the function to be estimated by a fuzzy neural network. Hence (1) can be rewritten as

$$z(t+1) = \bar{f}(z(t), z(t-1), \cdots, z(t-n), u(t), u(t-1), \cdots, u(t-m)). \tag{4}$$

Our purpose is to overcome skewness noises and to find a suitable identification model

$$\hat{z}(t+1) = \hat{f}(z(t), z(t-1), \cdots, z(t-n), u(t), u(t-1), \cdots, u(t-m)), \tag{5}$$

where $\hat{z}(t+1)$ is the output of the fuzzy neural networks and $\hat{f}$ is the estimate of $\bar{f}$. $\hat{y}(t+1)$ is the inverse of Box-Cox transformation $\hat{z}(t+1)$. That is,

$$\hat{y}(t+1) = BCT^{-1}(\hat{z}(t+1)) = \sqrt[\lambda]{1 + \lambda \tilde{y}^{\lambda-1} \hat{f}}. \tag{6}$$

The structure of radial basis function fuzzy neural networks consists of an input layer, a hidden layer of radial basis functions and a linear output layer. When the radial basis functions are chosen as Gaussian functions, a radial basis function fuzzy neural networks can be expressed in the form

$$\hat{z}_s(t+1) = \sum_{j=1}^{L} w_{js} G_j = \sum_{j=1}^{L} w_{js} \exp(-\frac{\|\vec{x} - m_j\|^2}{2\sigma_j^2}), \text{ for } s = 1, 2, \cdots, p, \tag{7}$$

where $\hat{z}_s$ is the sth output, $\vec{x} = (z(t), z(t-1), \cdots, z(t-n), u(t), u(t-1), \cdots, u(t-m))$ is the input to the fuzzy neural networks, $w_{js}, 1 \leq j \leq L, 1 \leq s \leq p$, are the synaptic weights, $G_j, 1 \leq j \leq L$, are the Gaussian functions, $m_j, 1 \leq j \leq L$, and $\sigma_j, 1 \leq j \leq L$, are the centers and the widths of G_j, respectively, and L is the number of the Gaussian functions, meanwhile, one can find that L also denotes the number of rules in the Box-Cox transformation-based ARFNNs.

3   The Main Results

The fuzzy inference systems has been proved to be a universal approximator [10], that is, they can approximate any real continuous function on a compact set to an arbitrary precision, provided sufficient fuzzy logic rules are available. The fuzzy rule base comprises the following fuzzy IF-THEN rules:

Rule i:

$$\text{IF } x_1 \text{ is } A_1^i \text{ and ... and } x_n \text{ is } A_n^i ,$$

$$\text{THEN } \hat{z}_1 \text{ is } B_1^i \text{ and ... and } \hat{z}_p \text{ is } B_p^i , \tag{8}$$

where A_j^i and B_j^i are fuzzy sets in $U_i \subset R$ and $V \subset R$, respectively, and $\vec{x}_i \in R^n$ and $z \in V$ are the input and output variables of the fuzzy system, respectively. Given a pair $(\vec{x}_i , z_i)$, the final output of the fuzzy system is inferred as follows:

$$\hat{z}_s = \frac{\sum\limits_{i=1}^{m} \beta_i (\prod\limits_{j=1}^{n} \mu_{A_j^i}(x_j))}{\sum\limits_{i=1}^{m} \prod\limits_{j=1}^{n} \mu_{A_j^i}(x_j)}, \quad s = 1,2,\cdots,p. \tag{9}$$

The proposed fuzzy neural networks consist of m rules in the form of (8) and that membership functions are Gaussian function. Besides, all variance in the membership functions are initalt set ia equal. Hence, the (9) can be rewrite as

$$\hat{z}_s(\vec{x}) = \sum_{i=1}^{m} W_i \Phi_i = \sum_{i=1}^{m} W_i \exp(-\frac{\left\| \vec{x} - \vec{x}_i \right\|^2}{2\sigma_i^2}) . \tag{10}$$

It is shown that fuzzy neural networks in (10) can be represented as the functional link networks that are based on Gaussian function. Besides, the proposed structure is shown in Figure 1, which is comprised by the input layer, membership functions layer, rule inference system, and output layer.

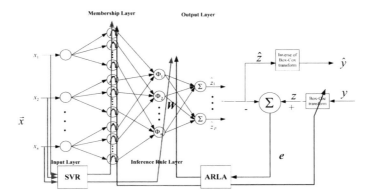

Fig. 1. The structure of proposed the Box-Cox transformation-based ARFNNs

An SVR approach is used to approximate an unknown function from a set of samples $\{(\vec{x}_i, z_i), i = 1, 2, \cdots, N\}$, where the system output y_i is replaced by the Box-Cox transformed response z_i. The problem is then be transformed into finding the parameters of the following basis linear expansion

$$f(\vec{x}, \vec{\theta}) = \sum_{k=1}^{m} \theta_k g_k(\vec{x}) + b, \tag{11}$$

where $\vec{\theta} \in (\theta_1, \ldots, \theta_m)$ is a parameter vector to be identified and b is a constant.

Vapnik [11] proposed ε-SVR approach in 1995, the solution for the problem is to find f that minimizes

$$R(\vec{\theta}) = \frac{1}{N} \sum_{i=1}^{N} L_\varepsilon (z_i - f(\vec{x}_i, \vec{\theta})), \tag{12}$$

subject to the constraint

$$\|\vec{\theta}\|^2 < C, \tag{13}$$

where $L_\varepsilon(\cdot)$ is the ε-insensitive loss function and defined as

$$L_\varepsilon(e) = \begin{cases} 0, & \text{for } |e| \le \varepsilon, \\ |e| - \varepsilon, & \text{otherwise,} \end{cases} \tag{14}$$

for some previously chosen nonnegative number ε, and C is a constant. In (13), the constraint is imposed to trade off the complexity of the solution. By using the Lagrange multiplier method, it can be shown that the minimization of (12) leads to the following dual optimization problem,

$$\text{Minimize } Q(\alpha, \alpha^*) = \varepsilon \sum_{r=1}^{N} (\alpha_r + \alpha_r^*) - \sum_{r=1}^{N} y_r (\alpha_r^* - \alpha_r)$$
$$+ \frac{1}{2} \sum_{r,s=1}^{N} (\alpha_r^* - \alpha_r)(\alpha_s^* - \alpha_s) K(\vec{x}_r, \vec{x}_s). \tag{15}$$

It was shown in [12] that the SVR uses the quadratic programming optimization to determine in the form of the following linear expansion of kernel functions

$$f(\vec{x}, \alpha, \alpha^*) = \sum_{k=1}^{m} (\alpha_k^* - \alpha_k) K(\vec{x}_r, \vec{x}_s) + b. \tag{16}$$

In this paper, the Gaussian function is used as the kernel function. Hence, (16) can be rewritten as

$$f(\vec{x}, \vec{v}) = \sum_{k=1}^{SV} v_k \exp(-\frac{\|\vec{x} - \vec{x}_k\|^2}{2\sigma_k^2}) + b, \tag{17}$$

where $v_k = (\alpha_k^* - \alpha_k) \neq 0$ and $\bar{x}_k$ are SVs. And let $\exp(-\frac{\|\bar{x} - \bar{x}_1\|^2}{2\sigma_1^2}) = 1, v_1 = b$.

Hence, (17) can be rewritten as

$$f_s(\bar{x}, v) = \sum_{k=1}^{SV+1} v_k \exp(-\frac{\|\bar{x} - \bar{x}_k\|^2}{2\sigma_k^2}). \tag{18}$$

From (10) and (18), the $i, m, W_i, \bar{x}_i, \sigma_i$, are equal to $k, (SV+1), v_k, \bar{x}_k, \sigma_k$, respectively. That is,

$$\hat{z}_s(\bar{x}) = f_s(\bar{x}, W) = \sum_{i=1}^{m} W_i \exp(-\frac{\|\bar{x} - \bar{x}_i\|^2}{2\sigma_i^2}). \tag{19}$$

The initial weights W_i and the number of rule m of the proposed Box-Cox transformation based ARFNNs can be determined via the SVR. Hence, we use the SVR to determine the initial structure for the fuzzy neural networks.

In this paper, we proposed ARLA to update the synaptic weights W_i, the centers $\bar{x}_i$, and the width σ_i of Gaussian kernels in the fuzzy neural networks, which uses the annealing concept in the cost function of robust back-propagation learning algorithm [6], can overcome the existing problems in robust back-propagation learning algorithm when system contain skewness noises. A cost function for the ARLA is defined here as

$$E_N(h) = \frac{1}{N} \sum_{i=1}^{N} \rho[e_i(h); \beta(h)], \tag{20}$$

where $e_i(h) = z(i) - \hat{z}_s(\bar{x}_i) = z(i) - \sum_{k=1}^{m} W_{ki} \exp(-\frac{\|\bar{x}_i - \bar{x}_k^i\|^2}{2\sigma_{ki}^2})$, $\tag{21}$

h is the epoch number, $e_i(h)$ is the error between the ith desired output and the ith output of the fuzzy neural networks at epoch h, the $\bar{x}_i$ is the ith input sampling value, the $\bar{x}_k^i$ and σ_{ki}^2 is the kth center and variance of membership function, $\beta(h)$ is a deterministic annealing schedule acting like the cut-off points and $\rho(\cdot)$ is a logistic loss function. Given suitable $\hat{\lambda}$, a maximum likelihood parameter estimator can be obtained by assuming that there exists a suitable Box-Cox transformation-based ARFNNs such that the model error $e_i(h) \sim N(0, \sigma^2)$ [13]. Support that there is a suitable Box-Cox transformation given by (3) such that the transformed system output z satisfies the normal assumption with probability density function [7] in relation to the original output y_i, $i = 1, 2, \cdots, N$ as

$$\frac{1}{\sqrt{2\pi}\sigma} \exp\left(-\frac{e_i(h)}{2\sigma^2}\right) J_i(\lambda), \tag{22}$$

where $J_i(\lambda)$ is the Jacobain of the Box-Cox transformation given by [14]

$$J_i(\lambda) = \frac{\partial z(y_i, \lambda)}{\partial y_i} = (\frac{y_i}{\tilde{y}})^{\lambda-1} \; i = 1,2,\cdots, N. \tag{23}$$

Define a log-likelihood function can be found in [13]. The maximum likelihood estimator can be solved by nonlinear least squares algorithm.

4 Simulation Results

The simulations were conducted in the *Matlab* environment. The root mean square error (RMSE) of the testing is used to measure the performance of the learned networks.

Example : A function in [15] with skewness noises ξ is defined as

$$y(x) = \exp(\frac{2\sin x}{x} + \xi), \; -10 \le x \le 10. \tag{24}$$

Two hundred training data $y(x)$ were generated by using uniformly distributed random $x \in [-10,10]$. Besides, the skewness noises are produced with uniform random numbers and chosen from the interval $(0,0.2)$. The initial RMSE of two

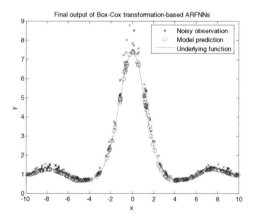

Fig. 2. The original data with skewness noises testing data (Initial RMSE = 0.3051).The final output for the proposed structure after 115 epochs uses the ARLA for Example (RMSE=0.0492).

hundred training data is 0.3051. Firstly, the training data use Box-Cox transformation by $\lambda = 0.7$ Assume that the membership function is Gaussian function and initial structures are obtained by an SVR approach. The parameters in SVR are set as $C = 1$, Gaussian kernel function with $\sigma = 0.75$ and $\varepsilon = 0.1$. The initial structure of the Box-Cox transformation -based ARFNNs with the number of rule in the simplified fuzzy inference system is obtained as 68. The parameters of the Box-Cox transformation-based ARFNNs are adjusted by ARLA. After 115 epochs using ARLA, the testing RMSE of the Box-Cox transformation-based ARFNNs is 0.0492, as shown in Figure 2. For a comparison study, the Box-Cox transformation-based RBFNs [7] are constructed for the same data, but the testing RMSE is 0.3079. Simulation results show that the Box–Cox transformation-based ARFNNs have fast convergent speed and more generalization capability for the function approximated problem with skewness noises.

Acknowledgments. This work was supported by the National Science Council of Taiwan, R. O. C. under Grand NSC 99-2221-E150-079 and National Formosa University EN99B-B1005.

References

1. Box, G.E.P., Cox, D.R.: An Analysis of Transformation. Journal of the Royal Statistical Society, Series B 26, 211–252 (1964)
2. Purwar, S., Kar, I.N., Jha, A.N.: On-line System Identification of Complex Systems Using Chebyshev Neural Networks. Applied Soft Computing 7, 364–372 (2007)
3. Trabelsi, A., Lafont, F., Kamoun, M., Enea, G.: Fuzzy Identification of a Greenhouse. Applied Soft Computing 7, 1092–1101 (2007)
4. Jeng, J.T., Chuang, C.C.: New Fuzzy Modeling Based on Input-Output Pseudolinearization and Its Digital Approximation Via Walsh Functions. International Journal of Fuzzy Systems 3, 503–511 (2001)
5. Leu, Y.G., Lee, T.T., Wang, W.Y.: On-line Tuning of Fuzzy-Neural Network for Adaptive Control of Nonlinear Dynamical Systems. IEEE Transactions on Systems Man and Cybernet 27, 1034–1043 (1997)
6. Chuang, C.C., Su, S.F., Hsiao, C.C.: The Annealing Robust Backpropagation (BP) Learning Algorithm. IEEE Transactions on Neural Networks 11, 1067–1077 (2000)
7. Hong, X.: A Fast Identification Algorithm for Box-Cox Transformation Based Radial Basis Function Neural Network. IEEE Transactions on Neural Networks 17, 1064–1069 (2006)
8. Hong, Y.P., Pan, C.T.: Rank-revealing QR Factorizations and the Singular Value Decomposition. Mathematics of Computation 58, 213–232 (1992)
9. Jeng, J.T., Chuang, C.C.: Selection of Initial Structures with Support Vector Regression for Fuzzy Neural Networks. International Journal of Fuzzy Systems 6, 63–70 (2004)
10. Jang, J.S., Sun, C.T., Mizutani, E.: Neuro-Fuzzy and Soft Computing. Prentice-Hall, Englewood Cliffs (1997)
11. Vapnik, V.N.: The Nature of Statistical Learning Theory. Springer, Berlin (1995)
12. Vapnik, V.N., Golowich, S., Smola, A.J.: Support Vector Method for Function Approximation, Regression Estimation and Signal Processing. Neural Information Processings Systems 9, 281–287 (1997)

13. Liu, Y.S., Su, S.F., Chuang, C.C., Jeng, J.T.: Box-Cox Transformation-based Annealing Robust Radial Basis Function Networks for Skewness Noises. In: 2010 International Conference on System Science and Engineering (ICSSE), pp. 537–541 (2010)
14. Chuang, C.C., Jeng, J.T., Lin, P.T.: Annealing Robust Radial Basis Function Networks for Function Approximation with Outliers. Neurocomputing 56, 123–139 (2004)
15. Hong, X.: Modified Radial Basis Function Neural Network Using Output Transformation. IET Control Theory & Applications 1, 1–8 (2007)

A Fuzzy PID Controller Based on Hybrid Optimization Approach for an Overhead Crane

Chia-Nan Ko

Department of Automation Engineering, Nan-Kai University of Technology
Tasotun, Nantou 542, Taiwan
t105@nkut.edu.tw

Abstract. A fuzzy PID controller is proposed to asymptotically stabilize a three-dimensional overhead crane using a hybrid optimization approach in this article. In the proposed algorithm, the PID gains are adaptive then the fuzzy PID controller has more flexibility and capability than the conventional ones with fixed gains. To tune the fuzzy PID controller simultaneously, a hybrid optimization procedure integrating genetic algorithm (GA) and particle swarm optimization (PSO) method is adopted. The simulation results illustrate that the proposed controller with few fuzzy rules can effectively perform the asymptotical stability of the prototype overhead crane.

Keywords: Three-dimensional overhead crane, particle swarm optimization, genetic algorithm, fuzzy PID controller, hybrid optimization approach.

1 Introduction

Overhead cranes have been widely used in industry for transportation systems. However, the overhead cranes have several problems. Such that load swing usually degrades work efficiency and sometimes causes load damages and even safety accidents in the worst cases. Therefore, some researchers have endeavored to control the load swing [1-8]. Most industrial processes nowadays are still controlled by PID controllers [9-12]. However, a conventional PID controller may have poor control performance for nonlinear and/or complex systems that have no precise mathematical models. The main disadvantage is that they usually lack in flexibility and capability.

Fuzzy controllers provide reasonable and effective alternatives for conventional controllers. Many researchers attempted to combine conventional PID controllers with fuzzy logic [13, 14]. Despite the significant improvement of these fuzzy PID controllers over their classical counterparts, it should be noted that they still have disadvantages. How to reduce the number of fuzzy rules is arduous work for nonlinear multivariable systems.

Several evolutionary algorithms have been proposed recently to search for optimal PID controllers. Among them, genetic algorithm (GA) has received great attention and particle swarm optimization (PSO) method has been successfully applied to various fields [15, 16]. In this paper, a hybrid optimization approach integrating GA and PSO will be adopted to perform the fuzzy PID control. To show the flexibility and capability of the proposed method, an overhead crane is adopted as an illustrative

T.-H.S. Li et al. (Eds.): FIRA 2011, CCIS 212, pp. 202–209, 2011.
© Springer-Verlag Berlin Heidelberg 2011

example. From the simulation results, one can find that the designed fuzzy PID controller guarantees not only prompt damping of load swing but also accurate control of crane positions.

2 Fuzzy PID Controllers

In the proposed fuzzy PID controller, the input variables of the fuzzy rules are the error signals and their derivatives, while the output variables are the PID gains. The fuzzy PID control rules are expressed as

If e_1 is X_1^i and $\dot{e}_1$ is X_2^j and e_2 is X_3^k and $\dot{e}_2$ is X_4^l,

then $K_{P1} = Y_{P1}^{ijkl}, K_{I1} = Y_{I1}^{ijkl}, \cdots, K_{D2} = Y_{D2}^{ijkl}$

for $1 \le i \le n_1, \ 1 \le j \le n_2, \ 1 \le k \le n_3, \ 1 \le l \le n_4$, $\qquad$ (1)

where e_1, e_2 and $\dot{e}_1$, $\dot{e}_2$ are the error signals and their derivatives, X_1^i, X_2^j, X_3^k, X_4^l are the membership functions of e_1, $\dot{e}_1$, e_2, and $\dot{e}_2$, $K_{P1}, K_{I1}, \cdots, K_{D2}$ are the PID gains, $Y_{P1}^{ijkl}, Y_{I1}^{ijkl}, \cdots, Y_{D2}^{ijkl}$ are real numbers, n_1, n_2, n_3, and n_4 denote the numbers of input membership functions, respectively.

The membership functions of a fuzzy system are usually parametric functions such as triangular functions, trapezoidal functions, Gaussian functions, and singletons. Though the proposed method is equally applicable to all these kinds of membership functions, asymmetric Gaussian ones are used as the antecedent fuzzy sets in this paper. This means that input membership functions are represented as

$$X_k^{m_i}(x_k) = \begin{cases} \exp\left[-\left(\dfrac{x_k - \rho_k^{m_i}}{\sigma_{kl}^{m_i}}\right)^2\right] & \text{if } x_k \le \rho_k^{m_i} \\[2em] \exp\left[-\left(\dfrac{x_k - \rho_k^{m_i}}{\sigma_{kr}^{m_i}}\right)^2\right] & \text{if } x_k > \rho_k^{m_i} \end{cases}$$

for $k = 1, 2, \cdots, 4, \ 1 \le m_1 \le n_1, 1 \le m_2 \le n_2, 1 \le m_3 \le n_3, 1 \le m_4 \le n_4$, $\qquad$ (2)

where x_k represents the input linguistic variables, $\rho_k^{m_k}$, $\sigma_{kl}^{m_i}$, and $\sigma_{kr}^{m_i}$ denote the values of the centers, the left widths, and the right widths of the input membership functions, respectively. For the output membership functions, singleton sets are adopted. In the defuzzification process, Wang [17] used the center of gravity method to determine the output crisp values. Then, if the PID control law is used and the control signal is determined as

$$u(t) = K_{P1}e_1(t) + K_{I1}\int e_1(t)dt + K_{D1}\dot{e}_1(t) + K_{P2}e_2(t) + K_{I2}\int e_2(t)dt + K_{D2}\dot{e}_2(t) \quad (3)$$

From the above description, one can find that the gains of the fuzzy PID controller are adaptive such that the controller should have more flexibility and capability than the conventional ones. However, it is very difficult, if not impossible, to determine the parameters directly. Therefore, a hybrid algorithm integrating PSO and GA is proposed to search for the optimal values of these parameters simultaneously.

3 A Simulation Example

3.1 Dynamic of an Overhead Crane

In practice, load swing is suppressed as much as possible for safety considerations. This study considers this practical case of small load swing around the stable equilibrium. Then, for the generalized coordinates x, θ_x y, and θ_y, in Fig. 1 the following linearized dynamic model [5] can be derived:

$$(M_X + m)\ddot{x} + mL\ddot{\theta}_x = F_x - D_x \dot{x}, \tag{4}$$

$$\ddot{x} + L\ddot{\theta}_x + g\theta_x = 0, \tag{5}$$

$$(M_Y + m)\ddot{y} + mL\ddot{\theta}_y = F_y - D_y \dot{y}, \tag{6}$$

$$\ddot{y} + L\ddot{\theta}_y + g\theta_y = 0, \tag{7}$$

where m is the load mass; L is the rope length; M_X and M_Y are the x and y components of the crane mass including the moment of inertia of the gear train and motors, respectively; D_X and D_Y denote the viscous damping coefficients of the crane in the x and y directions, respectively; F_x and F_y are the force inputs to the crane in the x and y directions, respectively; g denotes the gravitational acceleration.

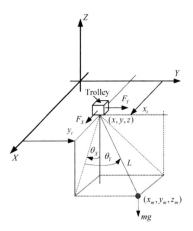

Fig. 1. Coordinate systems of an overhead crane

3.2 GA-PSO Tuning Fuzzy PID Controller

In the overhead crane, the desired value of $x(t)$ and $\theta(t)$ are denoted by x_d and θ_d. If the PID control law is employed, then the input-output relation of the crane system is expressed as

$$\tau(t) = k_{P1}e_1(t) + k_{I1}\int e_1(t)dt + k_{D1}\dot{e}_1(t) + k_{P2}e_2(t) + k_{I2}\int e_2(t)dt + k_{D2}\dot{e}_2(t), \qquad (8)$$

where $e_1(t) = x(t) - x_d$, $e_2(t) = \theta(t) - \theta_d$, $\dot{e}_1(t) = \dot{x}(t) - \dot{x}_d$, and $\dot{e}_2(t) = \dot{\theta}(t) - \dot{\theta}_d$.

3.3 Fitness

In designing the fuzzy PID controller, the primary goal is to drive an overhead crane system from the given initial state to the desired final state. However, if the number of fuzzy rules is large, then heavy computation burden and huge memory requirement are inevitable. Therefore, the primary goal and the way to reduce the number of fuzzy rules should be taken into account simultaneously in defining the fitness function. This means that two performance criteria are chosen as follows:

$$f = \frac{1}{[(1 + \sum_{i=1}^{n_1} p_i) \cdot (1 + \sum_{j=1}^{n_2} p_j) \cdot (1 + \sum_{k=1}^{n_3} p_k) \cdot (1 + \sum_{l=1}^{n_4} p_l)]^2} + \frac{1}{\int t[e_1^2(t) + e_2^2(t)]dt}, \qquad (9)$$

where p_i, p_j, p_k, and p_l, are the binary elements to indicate which ones of the membership functions are activated. From the definition (9), the fitness value can be calculated to evaluate the performance of the fuzzy PID controller and a higher fitness value denotes a better performance.

4 Integration of PSO and GA

PSO is a population-based stochastic searching technique developed by Kennedy and Eberhart [18]. It is similar to the GA in that it begins with a random population matrix and searches for the optima by updating generations.

4.1 Particle Representations

Before applying the novel auto-tuning method, how to encode the parameters must be introduced firstly. In the proposed method, a mixed coding method is used, in which n_1, n_2, n_3, and n_4 are encoded as binary numbers and $\rho_k^{m_i}$, $\sigma_{kl}^{m_i}$, $\sigma_{kr}^{m_i}$, Y_{P1}^{ijkl}, Y_{I1}^{ijkl}, Y_{D1}^{ijkl}, Y_{P2}^{ijkl}, Y_{I2}^{ijkl}, Y_{D2}^{ijkl} are encoded as real numbers. This means that the positions of particles are represented as

$$P = [p_{binary} \quad p_{real}]. \qquad (10)$$

The particle p_{binary} contains binary variables taking the value of one or zero. The elements of p_{binary} are used to indicate which ones of the membership functions are activated. As for the real particles p_{real}, the elements of p_{real} are used to represent the values of $\rho_k^{m_i}$, $\sigma_{kl}^{m_i}$, $\sigma_{kr}^{m_i}$, Y_{P1}^{ijkl}, Y_{I1}^{ijkl}, Y_{D1}^{ijkl}, Y_{P2}^{ijkl}, Y_{I2}^{ijkl}, and Y_{D2}^{ijkl}.

4.2 Evolutionary Algorithms

In evolutionary strategies, the real particles p_{real} will employ the PSO method. As for binary particles p_{binary}, it will adopt the GA because of their nature and simplicity. In PSO method, the particles update their velocities and positions based on the local best and global best solutions [19]. In the evolutionary procedure, the inertia weight, the cognitive parameter, and the social parameter are linearly adaptable over the evolutionary procedure [19]. In the proposed GA-based method for binary particles, one cut-point crossover operator and single-point mutation operator will be employed [20].

5 Simulation Results

The parameters of the overhead crane system shown as Fig. 1 are chosen as [2]

$$M_X = 1440 \; kg, \, D_X = 480 \; N \cdot s \, / \, m,$$

$$M_Y = 110 \; kg, \, D_Y = 40 \; N \cdot s \, / \, m,$$

$$m = 10 \; kg, \, L = 1 \; m, \, g = 9.8 \; m \, / \, s^2,$$

and the constraints shown as

$$-4800 \; N \le F_X \le 4800 \; N, -200 \; N \le F_Y \le 200 \; N,$$
$$-\pi / 12 \text{ rad/s} \le \theta \le \pi / 12 \text{ rad/s } -\pi / 6 \text{ rad/s} \le \dot{\theta} \le \pi / 6 \text{ rad/s},$$
$$0 \; m \le x \le 5.5 \; m, \quad 0 \; m \le y \le 3.5 \; m,$$
$$-0.5 \; m \, / \, s \le \dot{x} \le 0.5 \; m \, / \, s, -2 \; m \Big/ s^2 \le \ddot{x} \le 2 \; m \Big/ s^2,$$
$$-0.3 \; m \, / \, s \le \dot{y} \le 0.3 \; m \, / \, s, -1.5 \; m \Big/ s^2 \le \ddot{y} \le 1.5 \; m \Big/ s^2.$$

The initial state and the desired final state of the overhead crane are $(x, y, \theta) = (1, 1, \pi / 18)$ and $(x, y, \theta) = (0, 0, 0)$. In the proposed algorithm, the population size, the maximal iteration number, the crossover rate, and mutation rate are chosen to be 40, 2000, 0.8, and 0.2, respectively. Moreover, it is assumed that the values of n_1, n_2, n_3, and n_4 are all chosen as five, and the singletons of the output linguistic variables are all chosen as real numbers. According to the procedure of the GA-PSO algorithm, the minimal fuzzy rules and the optimal membership functions of the input linguistic variables are determined. Moreover, the optimal values of p_{binary} and p_{real} can be determined. The former is found to be [0101001101010011011 10] and it means that only the membership functions X_1^2, X_1^4, X_2^2, X_2^3, X_2^5, X_3^2, X_3^5, X_4^1, X_4^3, and X_4^4 are activated. Meanwhile, this also means that there are 36 $(= 2 \times 3 \times 2 \times 3)$ fuzzy rules in the fuzzy PID controller. Since the number of fuzzy rules is reduced from 625 ($= 5^4$) to 36, the computation burden in implementation of this fuzzy PID controller will also be reduced significantly.

In Figs. 2 and 3, the simulation results illustrate that the proposed fuzzy PID controller can effectively complete the asymptotical stability of the prototype overhead crane.

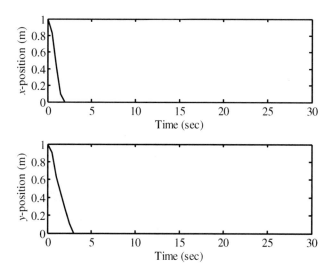

Fig. 2. Plots of x-position $x(t)$ and y-position $y(t)$ for the overhead crane using the proposed fuzzy PID controller

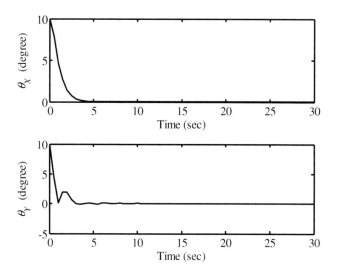

Fig. 3. Plots of X-angle $\theta_x(t)$ and Y-angle $\theta_y(t)$ of the overhead crane using the proposed fuzzy PID controller

6 Conclusions

In fuzzy PID tuning techniques, the parameters of fuzzy sets and PID gains are difficult to obtain the optimal values for stabilizing an overhead crane. In this paper, we present a hybrid optimization approach integrating GA and PSO to design a thoroughly self-tuning fuzzy PID controller to asymptotically stabilize the prototype overhead crane.

Acknowledgments. This work was supported in part by the National Science Council, Taiwan, R.O.C., under grant NSC 99-2221-E-252-017.

References

1. Yu, J., Lewis, F.L., Huang, T.: Nonlinear Feedback Control of a Gantry Crane. In: Proceedings of American Control Conference, pp. 4310–4315 (1995)
2. Lee, H.H.: Modeling and Control of a Three-dimensional Overhead Crane. J. Dyn. Systs., Meas., and Control 120, 471–476 (1998)
3. Singhose, W., Porter, L., Kenison, M., Kriikku, E.: Effects of Hoisting on the Input Shaping Control of Gantry Cranes. Control Eng. Pract. 8, 1159–1165 (2000)
4. Moustafa, K.A.F., Ebeid, A.M.: Nonlinear Modeling and Control of Overhead Crane Load Sway. J. Dyn. Syst., Meas., and Control 110, 266–271 (1988)
5. Cho, S.K., Lee, H.H.: A Fuzzy-logic Antiswing Controller for Three-dimensional Overhead Cranes. ISA Trans. 41, 235–243 (2002)
6. Hua, Y.J., Shine, Y.K.: Adaptive Coupling Control for Overhead Crane Systems. Mechatronics 17, 143–152 (2007)
7. Chang, C.Y.: Adaptive Fuzzy Controller of the Overhead Cranes with Nonlinear Disturbance. IEEE Trans. Ind. Inform. 3(2), 164–172 (2007)
8. Park, M.S., Chwa, D.Y., Hong, S.K.: Antisway Tracking Control of Overhead Cranes with System Uncertainty and Actuator Nonlinearity Using an Adaptive Fuzzy Sliding-mode Control. IEEE Trans. Ind. Electronics 55(11), 3972–3984 (2008)
9. Keel, L.H., Rego, J.I., Bhattacharyya, S.P.: A New Approach to Digital PID Controller Design. IEEE Trans. Automatic Control 48(4), 687–692 (2003)
10. Cervantes, I., Garrido, R., Jose, A.R., Martinez, A.: Vision-based PID Control of Planar Robots. IEEE Trans. Mechatronics 9(1), 132–136 (2004)
11. Whidborne, J.F., Istepanian, R.S.H.: Genetic Algorithm Approach to Designing Finite-precision Controller Structures. IEE Proceedings of Control Theory Applications 148(5), 377–382 (2001)
12. Lin, L., Jan, H.Y., Shieh, N.C.: GA-based Multiobjective PID Control for a Linear Brushless Dc Motor. IEEE Trans. Mechatronics 8(1), 56–65 (2003)
13. Tao, C.W., Taur, J.S.: Robust Fuzzy Control for a Plant with Fuzzy Linear Model. IEEE Trans. Fuzzy Syst. 13(1), 30–41 (2005)
14. Wu, C.J., Liu, G.Y., Cheng, M.Y., Lee, T.L.: A Neural-network-based Method for Fuzzy Parameter Tuning of PID Controllers. J. Chin. Inst. Eng. 25(3), 265–276 (2002)
15. Gaing, Z.L.: A Particle Swarm Optimization Approach for Optimum Design of PID Controller in AVR System. IEEE Trans. Energy Conversion 19(2), 384–391 (2004)
16. Habib, S.J., Al-kazemi, B.S.: Comparative Study between the Internal Behavior of GA and PSO through Problem-specific Distance Functions. In: The 2005 IEEE Congress on Evolutionary Computation, pp. 2190–2195 (2005)

17. Wang, L.X.: A Course in Fuzzy Systems and Control. Prentice-Hall, New Jersey (1997)
18. Kennedy, J., Eberhart, R.: Particle Swarm Optimization. In: Proceedings of the IEEE International Conference on Neural Networks, pp. 1942–1948 (1995)
19. Ratnaweera, A., Halgamuge, S.K., Watson, C.: Self-organizing Hierarchical Particle Swarm Optimizer with Time-varying Acceleration Coefficients. IEEE Trans. on Evolutionary Comput. 8(3), 240–255 (2004)
20. Haupt, R.L., Haupt, S.E.: Practical Genetic Algorithms, 2nd edn. John Wiley & Sons, New York (2004)

Design of High Power Density DC-DC Converter for Robots

Kuo-Ching Tseng and Ming-Han Tsai

Department of Electronic Engineering,
National Kaohsiung First University of Science and Technology
{jerry,u9952812}@nkfust.edu.tw

Abstract. In order to make the robot lighter, minimize the size, and have good heat dissipation, the technique of high power density DC-DC converter design is necessary. It can minimize the internal circuit in the robots, improve heat dissipation, and have high efficiency. Therefore, adoption of a forward converter synchronous rectification is used. This technique can diminish the loss of diode forward so it is able to increase the conversion efficiency. The active clamp reset transformer is used on primary-side to make Transformer demagnetization. At the same time, the power MOSFET can achieve the goal of zero voltage switching function, improving the efficiency of the converter. Finally, the forward converter is made with operating frequency of 300k Hz and 100W output power, and the maximum efficiency is 92.4%.

Keywords: power density, converter, robot.

1 Introduction

As technology developed rapidly these days, the needs for automation projects grow, and that is the reason why robots research becomes so popular. The robots research can be classified into various groups, such as industrial robots, guide robots, space exploration robots, household robots, and etc. The stability of robots [1-2] which has good power converter design must be improved. The power supplies heat will increase and it would lead to decline in efficiency. This paper discusses a high power density DC-DC converter application of planar transformer [3-4] and Development of a synchronous rectifier forward converter. The output power is 100W. To achieve reduced volume and power density enhance results [5-6].

2 Miniaturization of DC-DC Converter on Application of Planar Transformer

Conventional wire wound transformer is increasingly not applicable in DC-DC converter. Therefore, planar transformer gradually catches people's attention.

2.1 Design of Conventional Wire Wound Transformer

Fig.1 is the forward converter transformer design flow chart. Before the transformer starts to design, it need to set input voltage rang, output voltage, output power, operation frequency and maximum duty cycle of parameters converter.

T.-H.S. Li et al. (Eds.): FIRA 2011, CCIS 212, pp. 210–217, 2011.
© Springer-Verlag Berlin Heidelberg 2011

(a) Decision ΔB

In the choice of core when there is a need to pay attention to the magnetic flux density and temperature rise. It needs to choose the high permeability, low core loss high saturation flux density of the material, and saturation flux density is defined as Bs.

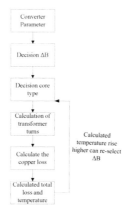

Fig. 1. Forward converter transformer design flow chart

Residual magnetic flux density is defined as Br, the maximum magnetic flux density variation ΔB is

$$\Delta B < B_s - B_r \quad [\text{mT}] \tag{2-1}$$

Magnetic flux density without being saturated under consideration, ΔB selected as

$$\Delta B = (B_s - B_r) \times 75\% \quad [\text{mT}] \tag{2-2}$$

Converter turn on time is defined as Ton, number of turns is defined as N. cross-sectional area of core is defined as *Ae*. Then ΔB is :

$$\Delta B = \frac{V_{in} \cdot T_{on}}{A_e N} \times 10^{-8} \quad [\text{mT}] \tag{2-3}$$

(b) Decision core type

The output power P_O is an important parameter when select the type of core. From P_O, it can decide the apparent power of transformer P_t. If the transformer has the primary winding and secondary winding. The primary winding power is defined as P_{in}. The Secondary winding power is defined as P_O. Apparent power is :

$$P_t = P_{in} + P_o \quad [\text{W}] \tag{2-4}$$

After decision to apparent power of transformer, it may determine the area product A_p is

$$A_p = A_e \cdot W_a \quad [\text{cm}^4] \tag{2-5}$$

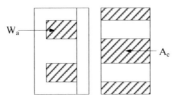

Fig. 2. Window area of core and cross-sectional area of core

Window area of core is W_a, cross-sectional area of core is A_e, as shown in Fig. 2 Apparent power can have conversion calculated from (2-5) :

$$A_p = \frac{p_t \times 10^4}{2\Delta BfJK_u} \quad [cm^4]$$

(2-6)

Therefore, decision A_p can search core type.

(c) Calculation of transformer turns
After design Maximum duty cycle D_{max} can obtain turns ratio as

$$n = \frac{N_p}{N_s} = \frac{V_{in(min)}D_{max}}{V_{out}}$$

(2-7)

By the turn ratio to calculate the primary winding turns is

$$N_p = \frac{V_{in(min)}T_{on}}{\Delta BA_e}$$

(2-8)

By primary winding turns can obtain secondary winding turns is

$$N_s = \frac{N_p}{n}$$

(2-9)

(d) Calculate the copper loss
Calculate the primary winding current Ip is

$$I_p = \frac{P_{in}}{\eta D_{max}V_{in}} \quad [A]$$

(2-10)

Where η is the efficiency of the converter, input current *RMS* is

$$I_{prms} = I_p \sqrt{D_{max}} \quad [A]$$

(2-11)

After calculating the primary side current *RMS*, it may assure that the secondary winding wire cross-sectional area can search wire's resistance R_{mean}. Transformer primary winding *MLT* (Length of Mean Turn) for MLT_{np} to the secondary winding resistance R_p as

$$R_p = N_p \times MLT_{np} \times R_{mean} \quad [\Omega]$$

(2-12)

When the output power is P_o, the secondary winding current can be obtained by approximation :

$$I_s = \frac{P_{out}}{V_o} \quad [A] \tag{2-13}$$

Secondary side current *RMS* is

$$I_{srms} = I_s \sqrt{D_{max}} \quad [A] \tag{2-14}$$

After calculating the secondary side current *RMS*, it may determine the secondary winding wire cross-sectional area can search wire mean resistance R_{mean}. Transformer secondary winding *MLT* for MLT_{ns} to the secondary winding resistance R_s as

$$R_s = N_s \times MLT_{ns} \times R_{mean} \quad [\Omega] \tag{2-15}$$

(e) Calculated total loss and temperature
Transformer copper loss is the primary side and secondary side of the copper loss sum :

$$P_{cu} = I_{prms}^2 R_p + I_{srms}^2 R_s \quad [W] \tag{2-16}$$

Search core Datasheet can obtain mean core loss P_{cv} and core volume V_e. Total core loss P_{cv} is

$$P_{fe} = P_{cv} \times V_e \quad [W] \tag{2-17}$$

The copper loss and core loss is transformer total loss P_{total} as

$$P_{total} = P_{fe} + P_{cu} \quad [W] \tag{2-18}$$

Search Datasheet can obtain surface area of core A_t, the surface area of transformer calculated wattage ψ is

$$\psi = \frac{P_{total}}{A_t} \quad \left[W\Big/cm^2 \right] \tag{2-19}$$

Transformer temperature rise T_r is calculated as

$$T_r = 450\psi^{0.826} \quad [^oC] \tag{2-20}$$

If the rise of calculated temperature is higher, it can re-select ΔB to adjust the primary and secondary winding turns.

2.2 Design of Planar Transformer

This section describes the design approach of planar transformer. Planar transformers can be constructed as stand-alone components, with a stacked layer design or a small multilayer PCB, or integrated into a multilayer board of the power supply. The advantages of planar transformer are: (a) low profile, (b) excellent thermal characteristics, (c) low leakage inductance, (d) excellent repeat ability of properties. Measurements on planar E core transformers under operating conditions with windings in multilayer PCBs show that the thermal resistance is substantially lower compared to conventional wire wound transformers with the same effective core volume. This is caused by the improved surface to volume ratio. The result of this better cooling capability is that planar transformers can handle higher throughput power densities, while the temperature rise is still within acceptable limits. In order to reduce the volume of converter, many engineers will use planar transformer windings in the main printed circuit board production.

3 Experimental Result

Fig.3 is structural figure of the planar transformer used. Layer 1 and Layer 4 is series. Layer 7 and Layer 10 is series. These two groups for parallel is secondary winding. Layer 2, Layer 3, Layer 5, Layer 6, Layer 8 and Layer 9 for series is primary winding. The transformer used 100 watts Synchronous Rectifier and active clamp DC-DC forward converter. Table.1 is Forward converter Parameter specification. Fig.4 is the thesis using the forward converter topology. Table.2 is planar transformer measured values.

Fig. 5(a) is output 100W and input $48V_{DC}$ when the voltage and current waveforms of primary side MOSFET Q_1. Oscilloscope channel 1(Ch1) is Voltage Stress waveforms of primary side MOSFET Q_1. Oscilloscope channel 3(Ch3) is current waveforms of primary side MOSFET Q_1. Oscilloscope channel 4(Ch4) is driving signal of primary side MOSFET Q_1.

Table 1. Forward converter Parameter specification

Project	Parameter specification
Input voltage rang	$36\ V_{DC}$~$75\ V_{DC}$
Normal input voltage	$48\ V_{DC}$
Output voltage	$5\ V_{DC}$
Output current	20 A
Output power	100 W
Operation frequency	300 kHz

Table 2. Planar transformer measured values

Project	Parameter specification
Primary winding turn	6 turn
Primary winding inductance	100.8 μH±5%
Secondary winding turn	2 turn
Secondary winding inductance	11.2 μH±5%
Primary winding resistance	8.4 mΩ±5%
Secondary winding resistance	2 mΩ±5%
Transformer size	31 mm × 26 mm × 8.3 mm

Fig. 5(b) is the voltage and current waveforms of primary side MOSFET Q_1. The channel 1(Ch1) is Voltage Stress waveforms of primary side MOSFET Q_1. The channel 3(Ch3) is current waveforms of primary side MOSFET Q_1. It can be seen when the MOSFET Q_1 conduction has reached the soft switch. Fig.6 shows the synchronous rectification MOSFET of driving signal waveform and the inductor current waveform.

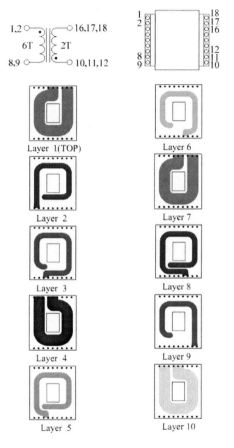

Fig. 3. Structural figure of the planar transformer used

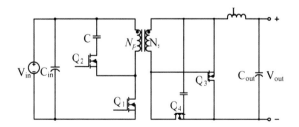

Fig. 4. The forward converter topology

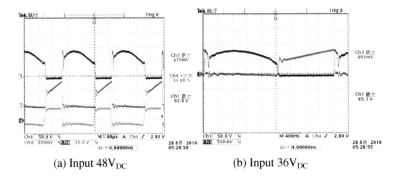

(a) Input 48V_DC   (b) Input 36V_DC

Fig. 5. Voltage and current waveforms of primary side MOSFET Q_1

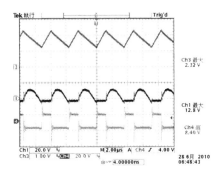

Fig. 6. Input 48V_DC when synchronous rectification MOSFET

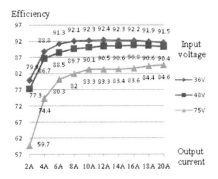

Fig. 7. Efficiency of converter

The channel 1 (Ch1) is the drive signal of Secondary side MOSFET Q_3. The channel 3 (Ch3) is secondary side inductor current waveform. The channel 4 (Ch4) is the drive signal of Secondary side MOSFET Q_4. Fig.7 is the efficiency of converter.

Conversion efficiency is about 92.4% of best. At 100W the efficiency of 90.4% in input 48V_DC. Output of 100W the efficiency of 84.6% in input 75V_DC. When the input voltage is 75V_DC, the duty cycle is only 22%; therefore less efficient.

Table 3. Temperature of component of converter

Component	Input 36 V_{DC} Temperature	Input 48 V_{DC} Temperature	Input 75 V_{DC} Temperature
planar transformer	81.7℃	88.9℃	92℃
MOSFET Q1	82.9℃	94.3℃	104.7℃
MOSFET Q3	69.5℃	72.8℃	80.7℃
MOSFET Q4	84.9℃	95.6℃	101.4℃

Table 3 shows $36V_{DC}$, $48V_{DC}$ and $75V_{DC}$ of the input voltages during 20 minutes after the temperature. Only when the input voltages are both 36 V_{DC} and $48V_{DC}$ will the temperature be in the acceptable range.

4 Conclusions

The design of internal power converter must be light, safe, energy-saving, and stable. As it becomes a trend, the size of DC-DC converters must be minimized. Miniaturization of components is important as well. Operating frequency of converter will increase. Using planar transformer to reduce the volume of transformer is necessary. This paper applied planar transformer to forward converter and use the active clamp and synchronous rectification circuit and to have good result in efficiency.

References

1. Anderson, I.A., Ieropoulos, I.A., McKay, T., O'Brien, B., Melhuish, C.: Power for Robotic Artificial Muscles. IEEE/ASME Transactions on Mechatronics 16(1) (February 2011)
2. Xue, L., Meyer, C., Dougherty, C.M., Bedair1, S., Morgan1, B., Arnold, D.P., Bashirullah, R.: Towards Miniature Step-Up Power Converters for Mobile Microsystems. In: Twenty-Sixth Annual IEEE Applied Power Electronics Conference and Exposition, APEC (2011)
3. Lim, B.S., Kim, H.J., Chung, W.S.: A Self-Driven Active Clamp Forward Converter Using the Auxiliary Winding of the Power Transformer. IEEE Trans. on Circuits and Systems 51(10), 549–551 (2004)
4. Senanayake, T., Ninomiya, T.: Auto-Reset Forward DC-DC Converter with Self-Driven Synchronous Rectification. In: IEEE PESC 2004, vol. 5, pp. 3636–3641 (2004)
5. Buccella, C., et al.: A Coupled Electrothermal Model for Planar Transformer Temperature Distribution Computation. IEEE Transactions on Industrial Electronics 55(10) (October 2008)
6. Han, Y., et al.: A Practical Copper Loss Measurement Method for the Planar Transformer in High-Frequency Switching Converters. IEEE Transactions on Industrial Electronics 54(4) (August 2007)

Applications of Fuel Cell Power Management System for Robot Vehicles

Kuo-Ching Tseng and Feng-Jie Chiou

Department of Electronic Engineering
National Kaohsiung First University of Science and Technology, Taiwan R.O.C.
{jerry,u9954603}@nkfust.edu.tw

Abstract. To improve stability of the robot power supply when perform task without voltage drop or sufficiency; therefore, a design of high-efficiency energy-saving method for power management is presented. It can be used in rechargeable lithium batteries and monitoring current output of different voltage so that the power contained inside the robot is able to be used for robot vehicles. By a micro-controller monitors the Robot vehicles use the CAN Bus protocol to receive the fuel cell internal parameter values and transmission power values of the robot vehicles, in the QGVA 5.7-inch touch screen display, so that drivers can immediately to control the operational status of the robot vehicles.

Keyword: Robot vehicle, fuel cell, power management system.

1 Introduction

As technology developed rapidly, the study of robot becomes popular. The function of the auto-controlled robot technology is to work, move, and track. In order to make the robot perform effectively on doing tasks, a modularized highly intelligent power management system of robot design is invented [1]. When the carbon Monoxide reduction gradually become the topic of interest, the transportation with environmental protection and pollution-free alternative energy sources will be of more development. Alternative energy sources commonly use rechargeable battery, solar energy or Robot vehicles. It can reduce carbon emissions, but the charging time is too long and it also has the time constraints of charging. The market is less common. Solar energy electric are as a whole expensive and limited to the use where there is only suitable in the sunlight to provide energy, so the development will be limited. Robot vehicles will be the future trends of environmental protection, Robot vehicles power management system will also be increasingly important. [2-4]

2 The Fuel Cell Power Management System for Robot Vehicles Profile

Fig.1. shows the applications of fuel cell power management robot vehicles system architecture diagram, the vehicle energy is provided by the fuel cell as the main

T.-H.S. Li et al. (Eds.): FIRA 2011, CCIS 212, pp. 218–225, 2011.

power, solar panels for auxiliary power. Power supply system using Fuel Cell System provides output power of 4kW, 60-116V output voltage as a major power, the output voltage of the fuel cell system can not directly drive three-phase AC induction motor and will change due to the load, the output voltage will also change, so the output voltage of the fuel cell system as a DC / DC Step-Up Converter input voltage boost converter allows the use of high voltage to 365V, and stabilize the output voltage, the output voltage will not change due to the load, the output voltage will also change, high-boost converter which consists of four 1kW DC / DC Step-Up Converter; DC / DC Step-Up Converter outputs voltage 365V to recharge LiFePO4 Battery, LiFePO4 Battery will provide the energy in the three-phase AC motor drive, three-phase AC motor drive AC induction motor drive, so the car can move. The cars use Solar Cell System for the auxiliary power output, Maximum Power Point Tracking (MPPT), so that the output power of Solar Cell System will get the maximum power, and charge the 12V LiFePO4 Battery. The energy of 12V LiFePO4 Battery will use different DC / DC Converter to raise or lower the voltage to provide the system.

Fig. 2. shows the use of proton exchange membrane fuel cell module, consists of multi-chip fuel cell battery, single fuel cell open circuit voltage is about 1V. When it's at full load, the voltage will drop to 0.6 V so the fuel cell module maximum output power is 4Kw. The battery voltage changes due to the load and the output voltage is about 60V ~ 116V. Table 1. is the proton exchange membrane fuel cell module specification sheet.

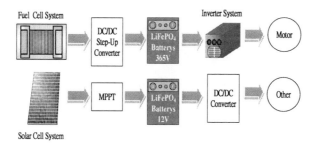

Fig. 1. Applications of fuel cell power management robot vehicles system architecture diagram

Fig. 2. Fuel cell module

Table 1. Fuel cell module specification

Fuel Cell Stack			
Rated Voltage	72V	Pressure of H_2	0.6 ~ 0.8bar
Rated Current	55A	Consumption of H_2	49L/min

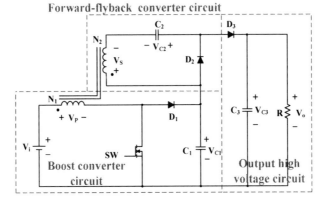

Fig. 3. DC / DC Step-Up Converter circuit

Table 2. DC / DC Step-Up Converter specification

Input Voltage	60V~116V	Frequency	33kHz
Output Voltage	365V	Output Power	1000W

The system provides the required output power of the vehicle, thus using the power converter output voltage to provide energy to the load stably. Fig. 3. shows the DC / DC Step-Up Converter circuit used in the robot vehicles [5]. Table 2. is the DC / DC Step-Up Converter specification sheet.

3 Software Design

To be able to immediately grasp the internal condition of fuel cell electric vehicle system, it uses Microchip's dsPIC30F4011 and PIC24FJ128GA006 [6-7]. Fig. 4 shows the chart of software design, as shown using dsPIC30F4011, to receive the information transmitted by CAN Bus, read A / D pin and the output signal source. Fig. 5(a). for the dsPIC30F4011 chip main program software design procedure, the program begins to be initialized to set Init_ ADC, Init_ UART and Init_ CAN. Fig. 5(b) .is the dsPIC30F4011 chip Timer1 interrupt subroutine procedure chart, when the Timer1 interrupts, it starts sampling AD, waiting to capture the required sampling, and begins to convert the signal, and the AD values will be stored in the temporary conversion.

Fig. 5(c). dsPIC30F4011 ingle-chip CAN Bus is the procedure chart of receiving interruption subroutine programs. When obtaining valid data of the fuel cell, it would identify the SID, process the data and store it in registers, clear the receiving interruption flags in CAN Bus, then finally end the interruption.

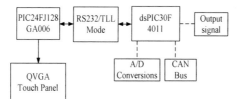

Fig. 4. Software design

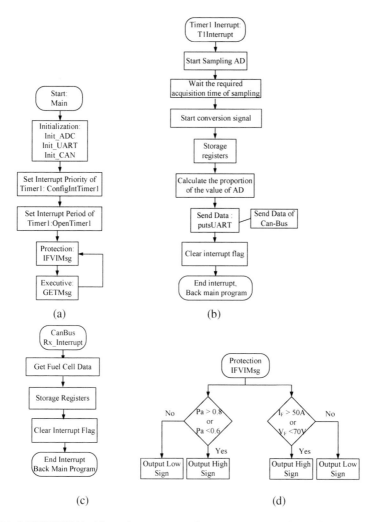

Fig. 5.(a) dsPIC30F4011 chip main program software design, (b) Timer1 interrupt subroutine procedure, (c) CAN Bus receive subroutine procedure, (d) Protection subroutine procedure.

Fig. 5.(d) shows the procedure chart of single-chip protection identification subroutine programs, the judging of protection circuit is to determine if the output voltage of fuel cell power system (V_F) is less than 70V, and output current (I_F) greater than 50A (Pa), and the output pressure greater than 0.8 bar or less than 0.6 bar. Fig. 6. is the procedure chart of the single-chip PIC24FJ128GA006 software design. The program begins to be initialized and create an object. At this moment it decides the number of objects, then draws the object, display pictures or text; detect whether there is contact with the touch screen, the change of voltage value. If it detects the touching of screen; it will perform the action; and create a new object, which also produce a new screen.

The action process of QVGA touch screen in software as shown in Fig. 7, the screen will stop at home page in the beginning. There are total of six objects, namely the fuel cell object, the object converter, the solar energy objects, the monitoring object, the object converter and the setting of the object. The system will detect whether something touches the button of objects. If there is a touch on the button of objects, it will enter the new page of the button of the objects which are touched. The new page displays information about the object. If you want to exchange for other page, just touch the touch screen, and it will jump back to the home page, previous or next page of the objects. The user may choose to return to the home page, the previous or next page of objects at will. The following is the monitoring system screen of design of fuel cell power supply system. Fig. 8. is the main screen control systems. The main monitor screen shows the six small controlled systems, the fuel cell power systems, high step-up DC converter system, solar energy auxiliary circuit system, the current monitoring system, the frequency convertor and battery, respectively.

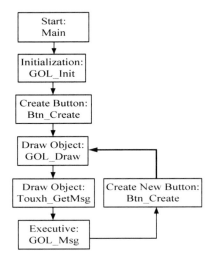

Fig. 6. PIC24FJ128GA006 software design

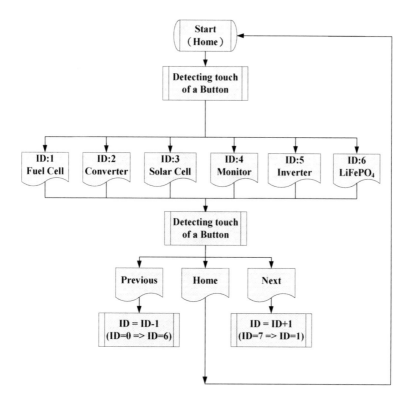

Fig. 7. The action process of QVGA touch screen in software

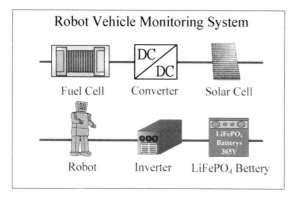

Fig. 8. Robot vehicles Monitoring System

4   Combination of Software and Hardware Test Results

Fig. 9. shows the main circuit detection and control structure. It will flow the hydrogen into fuel cell power supply system. Hydrogen output pressure control is between 0.6 ~ 0.8 bar. The internal status of fuel cell power supply system provides CAN Bus protocol, the internal current, such as output voltage (V_F), output current (I_F) and the battery temperature (T) state, and send the data via the CAN Bus, at this time, using a single chip dsPIC30f4011 to receive and processed the information sent by the fuel cell power system, and send to QGVA touch screen display. When the fuel cell output voltage (V_F) is too low, the output current (I_F) is too high, then sending the signal, Q_1 disconnect the fuel cell boost converter and high pressure rise convertor and make the protective circuit activate. To detect the converter voltage, current, and the LiFePO4 battery voltage, current will use single-chip dsPIC30f4011.

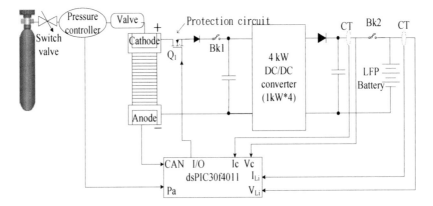

Fig. 9. Main circuit detection and control structure of the Robot vehicles system

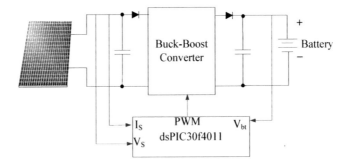

Fig. 10. Detection architecture of Solar Cell auxiliary circuit

Fig. 10. shows the detection architecture of solar panels auxiliary circuit. Detect the solar energy output voltage (V_S), output current (I_S) by dsPIC30f4011 IC, and for maximum power tracking (MPPT), control Buck-Boost converter.

5 Conclusions

As the technology grows faster and faster, robots must be safe, environment-friendly and energy-saving. As a result, the need of robots will be used available. In this paper QGVA touch screen display system, the internal current state of the fuel cell energy supply system uses of Microchip's single-chip as a data processing and transmission, so users can immediately grasp the electric power management system of the robot vehicles.

References

[1] Xie, L., Qiao, Q., Wang, Z.: An Efficient Power Management System for Biped Robot (October 2008)
[2] Cheng, Y.C.: fuel cell and electric vehicles. Science Development (367) (July 2003)
[3] Chen, C.X.: promote the application of a new era of hydrogen energy - Energy Office to promote hydrogen and fuel cell industry and technology. Ministry of Economic Affairs Bureau of Energy - Energy Report (June 2008)
[4] De Michieli, L., Nori1, F., Pini Prato, A., Sandini, G.: Study on Humanoid Robot Systems: an Energy Approach (August 2008)
[5] Tseng, K.C., Liang, T.J.: Novel high-efficiency step-up converter. IEE Proceedings Electric Power Applications 151, 182–190 (2004)
[6] PIC24FJ128GA006 datasheet
[7] dsPIC30F4011 datasheet

A Study on Control of a Small Fuel Cell Power Supply System for Robots

Kuo-Ching Tseng, Pao-Chuan Tseng, and Shih-Hsien Yang

Department of Electronic Engineering,
National Kaohsiung First University of Science and Technology
{jerry,u9752509,u9752004}@nkfust.edu.tw

Abstract. A hydrogen fuel cell power supply system which includes pressure valve, solenoid valves, flow controllers, temperature sensors and Proton Exchange Membrane Fuel Cell (PEMFC) is presented in the paper. The power supply system builds a control circuit loop includes a charge-discharge circuit with buck conversion topology. The feedback control signal of the power supply system by digital signal processor (dsPIC). This structure can achieve the goal of the small fuel cell power supply system stability, have high efficiency fuel cell, high efficiency converters, and to provide energy to a small robot.

Keywords: fuel cell, PEMFC, converter, digital signal processor.

1 Foreword

Today robot power supply system is provided by the battery. Because battery charging time is quite long, the energy robots cannot be supplied immediately. Therefore, the fuel cell plays a very important role. The way fuel cell generates electricity depends on hydrogen. If the hydrogen runs out, it can be filled immediately by adding the hydrogen bottle. Unlike a battery that cost a long time to recharge, filling hydrogen into the bottle just need little time. In the future hydrogen bottles will be possible to purchase in convenience stores, which can make the usage simpler.

About 98% nature resources used in Taiwan rely on importing and the petroleum occupies 50%. It becomes the biggest burden to Taiwan; hence, renewable energy development and energy management will be the main focus in the future. The coming of hydrogen-energy era of Taiwan will bring more benefits on economic development because the production of hydrogen is no longer be limited by a region but technique [3]. The reason why positive development is made on hydrogen's related energy science and technology are: high converting efficiency, zero pollution, easy access of hydrogen. Consequently, the fuel cell seems to be more important [3-4].

Nowadays many countries and professionals are gradually paying more attention on fuel cell, so researchers actively study to break several key techniques in recent years and makes the fuel cell techniques applied in many countries in daily lives [5-6].The fuel cell is a power supply equipment with high efficiency and low pollution. When the chemical energy is stored in oxidizing agent, this equipment can convert the chemical power into electric energy. The development have diverse prospect, like electricity used in families, transportation tool, and etc.

T.-H.S. Li et al. (Eds.): FIRA 2011, CCIS 212, pp. 226–232, 2011.
© Springer-Verlag Berlin Heidelberg 2011

This research use proton exchanges film fuel cell to convert hydrogen from chemical energy to electric energy. Its biggest advantage is that during respond process it doesn't pollute, and energy-converting efficiency is up to 60%~80%. In practical use efficiency is 2~3 times than that of traditional gas engine. Fuel cell has other advantages: the diversity on fuel source, clean exhaustion, low noise, little pollution on environment, high reliability, and etc [7-8]. The combination of device of Proton Exchange Membrane Fuel Cell, power converter, micro-controller, hydrogen-loop control will develop a set of intelligent system of the fuel cell power supply, and provide the stable electric power to a notebook.

2 Introduction of Fuel Cells

Now various types of fuel cells are presented. Because its operation temperature is different and it use different fuel, so it cause discrepancy on the electrolyte. Therefore, different types of fuel cells have different field to be applied [9].In this research it will mainly use small scaled robot to be the design principle. Low working temperature (30~80°C), higher power density, small size, light weight- these are the best qualities to put into proton Exchange Membrane Fuel Cell.

2.1 The Internal Structure of Fuel Cells

The proton exchanges film fuel cell is series-connected by many sets of single battery. Single battery can be simply divided into three parts: the Electrode, the Electrolyte Membrane, and the Bipolar Plate. Figure.1 shows the structure of PEMFC. The electrode is divided into anode and cathode; inside the structure there are the components of catalyst layers and diffusion layers. The function of catalyst layer is to catalyze the fuels cell to carry on electrochemistry reaction. There are two functions of the air-enlarged layer. First, it can make the air of reaction spread equally to the catalyst layer. Second, it should have the function to respond and Exhaust streaming way of bipolar Plate. The Membrane Electrode Assembly (MEA) is composed by three elements: anodes, Proton Exchange Membrane and cathodes. It is the central idea of PEMFC and is the main electrochemistry respond area for fuel cell. As a result, the capability of PEMFC depends on whether the performance of MEA is good [10].

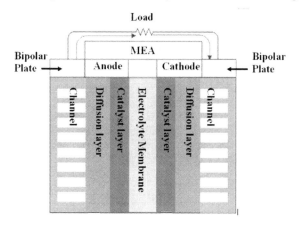

Fig. 1. PEMFCs constructs sketch map

3   The System of the Fuel Cell Power Supply

In this research, the power supply system uses the processor of the dsPIC digital signals of Microchip as the central control nucleus, and use the hydrogen loop to control the working temperature, adjust hydrogen barometric pressure, control hydrogen flow and electric energy loop. Fig.2 shows the system structure, including electric energy loop, buck converter, Microcontroller, the dsPIC 30 F4011 digital signal processor [11].

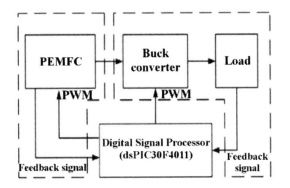

Fig. 2. System structure

3.1   Power Control Loop

The fuel cell output voltage is easily affected by the load fluctuation and polarized loss, which makes the DC outputted port voltage unsteady [12-13]. To conquer the problem, the power supply converting technique is adopted. In this research it uses the 30 W proton exchange film fuel cell and its output voltage is 11~21 Volt. As for the power, it is chosen to use which can be applied to 3C products, like Eee PC 701. The electric energy loop properly controls energy allocation and burden offset, provides stabilization output voltage, and raises the efficiency of whole power supply system. As fig. 3 shows, the control of power loop is to use digital signals processor that can carry

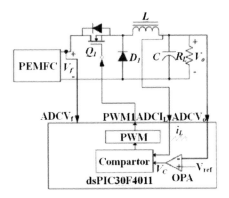

Fig. 3. Power control loop structure

out current mode control method [14]. When load fluctuates, output feedback voltage ADCV$_O$ and reference voltage V$_{ref}$ to zoom in and control voltage through the ratio of the error margin amplifier. Send out PWM signal through PWM generator to power switch, and control inductor current to adjust output voltage. In order to protect fuel cell, avoid excessive extraction of current, or instability of converter which may cause problems on power supply system, the design of detective output voltage V$_f$ of the fuel cell is needed. When V$_f$ is bigger than 11 Vs, use soft start to avoid surge wave current causing damage on fuel cell or power switch. Then carry on the control of the voltage mode to get stabilization of output voltage. When V$_f$ is smaller than 11 Vs, stop the converter and detect inductor current ADCI$_L$ to protect component and avoid damage [15-16].

3.2   Gas Loop Controls

Fig.4 shows the fuel cell air streaming road. It is divided into hydrogen and air loop, and takes hydrogen as the main air loop. The hydrogen is emitted through low-pressure hydrogen bottle. The electron press controller will adjust the fixed pressure, and mass controllers will adjust proper hydrogen flow. After that, deliver hydrogen to a gas hole of fuel cell, and to have electrochemistry respond with the oxygen of the air in the fuel cell pile. If fuel of hydrogen is not respond completely, it would deliver through an exhaustion hole, controlled by electromagnetism valve to press and leak pressure [17]. Pulse exhaustion is adopted in the research and exhaust per 15 seconds to raise the utilization rate of respond gas and save cost of fuel cell gas. The fan blows oxygen in the air to the battery pile through its gap. The fan adjusts the wind properly to get proper amount of oxygen to supply the fuel cell [18]. The control of hydrogen loop controls the oxygen flow and pressure properly to reduce fluctuation of the fuel cell gas, decrease the fluctuation of output voltage, increases electricity supply efficiency. It is controlled by the digital signals processor to control hydrogen loop. The steps are as follows:

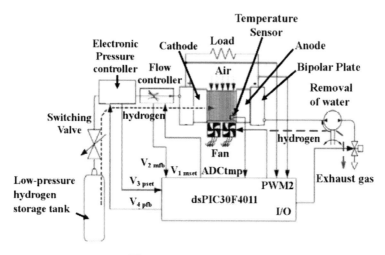

Fig. 4. Gas loop control structure

(1) Hydrogen flow control: The hydrogen flow supply varies as burden changes; the bigger the load, the more hydrogen flow will be. If the load is smaller, the hydrogen flow will become smaller. Therefore, it is needed to detect the loading of the current to control the hydrogen flow.

(2) Fan and working temperature control: When fuel cell sets generate electricity, as the burden enlarges the temperature will relatively raise, so the fan need to generate more wind to avoid excessive temperature, and increase the amount of oxygen in the air. So detect the current of fuel cell is necessary to adjust the fan voltage (PWM 2).

(3) Pressure control: When change the pressure of input hydrogen, the hydrogen flow will change as well. So it would detect the hydrogen flow, the fuel cell current, and its voltage, to decide the degree of adjustment on pressure.

4   Experimental Result

Fig.5 shows the efficiency of load output power from 9W to 27W, and the full load efficiency of power supply system is about 91.5%. Figure.6 is this power supply in load

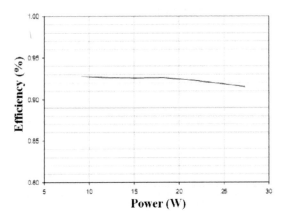

Fig. 5. The curves efficient of the overall power supply system

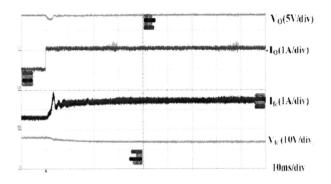

Fig. 6. The waveform variation of V_o, I_o, I_{fc} and V_{fc} with 1A to 2A load moment of change

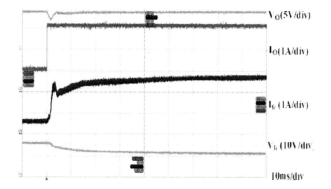

Fig. 7. The waveform variation of V_o, I_o, I_{fc} and V_{fc} with 1A to 3A load moment of change

of the power supply system to moment change 1-2 A. Fig.7 shows the moment change 1-3 A at the load for the power supply, the wave form of V_o, I_o, I_{fc}, and V_{fc}. The output voltage is steady in 9V in abrupt change of burden, the current of fuel cell rises suddenly, while fuel cell voltage decreases suddenly. As for the power loop, digital signals processor adjusts the duty of PWM signal to maintain stable output voltage. For hydrogen loop, when burden increase suddenly and the load current will increase suddenly from fuel cell. Then the rotational speed of fan will be adjusted gradually to raise the fuel cell voltage and drop off the temperature.

5 Conclusion

There are some features of this system structure. First, better than tradition rechargeable battery as the device of energy storage, if the sufficient fuel is provided, it can supply power to burden without any limitation. Second, because the microcontroller controls the hydrogen loop and electric energy loop, it can acquire some advantages of the power supply system of intelligent fuel cell, such as power stability, good performance of fuel cell, high efficiency on burden. Third, the fuel cell is developed by miniaturization big power, and it will be easy to carry. This research uses small scaled robot to be its loads device. In the future, the application of fuel cell will be widely and also bring infinite commercial potential.

References

[1] Kordesch, K., Simader, G.: Fuel Cells and Their Applications (January 1996)
[2] Wai, R.J., Lin, C.Y., Chu, C.C.: High step-up DC-DC converter for fuel cell generation system. In: Proc. IEEE IECON, pp. 57–62 (2004)
[3] Riezenman, M.J.: Fuel Cells for the Long Haul, Batteries for the Spurts. IEEE Trans. Power Electron 38, 95–97 (2001)
[4] Grove, W.R.: On voltaic series and the combination of gases by platinum. London and Edinburgh Philosophical Magazine and Journal of Science, Serirs 3, 127–130 (1839)

[5] Mond, L., Langer, C.: A new form of gas battery. Proceedings of the Royal Society of London, 296–204 (1889)

[6] Rahman, S.: Fuel cell as a distributed generation technology. In: Proc. IEEE PESS, pp. 551–552 (2001)

[7] Appleby, A.J., Foulkes, F.R.: Fucl Cell handbook. Van Nostrand Reinhold, New York (1989)

[8] Erickson, R.W., Maksimovic, D.: Fundamentals of Power Electronics, 2nd edn. Kluwer-Academic, Norwell (2001)

[9] Jin, K., Ruan, X., Yang, M., Xu, M.: A Novel Hybrid Fuel Cell Power System. In: Proc. IEEE PESC, pp. 1–7 (2006)

[10] Zhenhua, J., Dougal, A.R., Leonard, R.: A novel digital power controller for fuel cell/battery hybrid power sources. In: Proc. IEEE PESC, pp. 467–473 (2005)

[11] Qiao, H., Zhang, Y., Yao, Y., Wei, L.: Analysis of Buck-Boost Converter for Fuel Cell Electric Vehicles. In: IEEE ICVES, pp. 109–113 (2006)

[12] Thounthong, P., Sethakul, P.: Analysis of a Fuel Starvation Phenomenon of a PEM Fuel Cell. In: IEEE PCCON, pp. 731–738 (2007)

[13] Ticianlli, E.A., Derouin, C.R., Rdondo, A., Srinivasan, S.: Methods to Advance Technology of Proton Exchange Membrane Fuel Cell. Journal of Electrochemical Society 135, 2204–2214 (1999)

[14] Pukrushpan, J.T., Peng, H., Stefanopoulou, A.G.: Control -oriented modeling and analysis for automotive fuel cell systems. Transactions of the ASME, 14–25 (2004)

[15] Wang, L., Husar, A., Zhou, T., Liu, H.: A parametric study of PEM fuel Cell performance. International Journal of Hydrogen Energy, 1263–1272 (2003)

[16] Moore, R.M., Hauer, K.H., Friedman, D., Cunningham, J., Badrinarayanan, P., Ramaswamy, S., Eggert, A.: A dynamic simulation tool for hydrogen fuel cell vehicles. Journal of Power Sources, 272–285 (2005)

[17] Chen, F., Chu, H.S., Soong, C.Y., Yan, W.M.: Effective schemes to control the dynamic behavior of the water transport in the membrane of PEM fuel cell. Journal of Power Sources, 243–249 (2005)

[18] Eckl, R., Zehtner, W., Leu, C., Wagner, U.: Experimental analysis of water management in a self-humidifying polymer electrolyte fuel cell stack. Journal of Power Sources, 137–144 (2004)

Development of Simulator for
AndroSot in FIRA

Ping-Huan Kuo and Tzuu-Hseng S. Li

aiRobots Lab., Department of Electrical Engineering,
National Cheng Kung University,
No.1, University Road, Tainan City 701, Taiwan (R.O.C.)
coll22000@hotmail.com, thsli@mail.ncku.edu.tw

Abstract. This simulator is developed for Android Soccer Tournament (AndroSot) in FIRA. Due to the robot soccer game presents a dynamic and complex environment, it provides a challenging platform for multi-agent research. However, if there were some problems occurred in the robot actions and image processing algorithm, it is very difficult to run or test strategy systems. In order to solve these issues, a humanoid robot soccer competition's strategy simulation system is proposed, which provides developer to test the feasibility and advancement of the game strategy. In this simulator, strategies which compiled to DLL files may be explicitly loaded at run-time.

Keywords: simulator, AndroSot, FIRA.

1 Introduction

The robot soccer game system is a challenge for real-time control, which can be moderately abstracted from the standpoint of AI (Artificial Intelligence) and multi-agent systems. The robot soccer game is suitable for multi-agent system research, where the robots in one team have to cooperate in the face of competition with the opponent team. There is an international robot-soccer association, which is the FIRA Robot World Cup [1]. It started in 1996, together with many other FIRA events, will help generate interests in robotics in the young minds. However, it needs high threshold to participate in the competition. And the examination of its strategy is also troublesome.

In order to address these problems, development of simulator can be used for testing the feasibility and advancement of the game strategy of each team easily. Besides, for the FIRA SimuroSot game, the 3D two-wheel robot soccer simulator was developed by the RSS Development team and directed by Dr. Jun Jo of Griffith University, Australia. The aim of simulate competition is used for providing game training and strategy learning environment for each team. All in all, simulator provides a very convenient operating environment for the strategy testing of robot soccer competitions.

In terms of humanoid robot soccer, there are numerous competitions held around the world, and one of them is AndroSot. A humanoid robot belongs to highly intelligent system. The intelligent technologies of the humanoid robots include

T.-H.S. Li et al. (Eds.): FIRA 2011, CCIS 212, pp. 233–240, 2011.

mechanism design, vision system, and algorithms in software programming. Therefore, its competitions can encourage creativity and technical development. To facilitate the strategy testing of these humanoid robot soccer competitions, a strategy simulator for AndroSot in FIRA is proposed. Simulate competition software is compiled using Embarcadero C++ Builder 2010 [2], and it can run under Windows XP and Windows 7. In this simulator, strategies compiled to DLL files may be explicitly loaded at run-time. In addition, another simulation mode, "Master Level" is presented. The healthy state of each robot is considered in this mode. If the robot collided with another robots or the wall, its value of the healthy state will be reduced. If the value of healthy state is zero, the robot cannot execute any action.

2   System Architecture

First of all, the system flowchart of the simulator is shown in Fig. 1. When the "Start" button is clicked by the user, the background music composed by ourselves is played. Next, the simulator will check whether the DLL files are loaded or not. If the DLL strategy files are loaded, the simulator will execute the strategies in the DLL files. On the contrary, if the DLL strategy files are not loaded, the simulator will execute the built-in strategies. After that, the pose of the robots will be set immediately. Then, if the simulation level selected by the user is "Master level", the healthy point (HP) of each robot will be reduced or increased. The value of HP depends on the

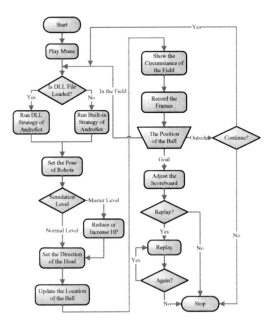

Fig. 1. System flowchart of the simulator. System process is running according to it.

circumstances during the competition. Afterward, the location of the ball is updated. After all of the processes described above are completed, in order to show the circumstances in the playing field, the simulator will use computer graphics to simulate the field, the robots and the ball. Moreover, for the purpose of replaying the process, the current frame will be recorded in the memory. Then, the position of the ball will be checked for the post-process. That is, if the ball is outside, the simulator will show the dialog box to confirm whether the competition should be continued or not. On the other hand, if the ball is in the goal, the scores on the scoreboard will be adjusted. Then the simulator will show the dialog box to confirm whether the user wants to replay or not. If the user does not want to replay, the simulation will be stopped. However, the user also can click the "Start" button to start the match again.

3   User Guide of the Proposed Simulator

Fig. 2 shows the user interface of robot soccer simulator. The robots are divided into two teams: team A (blue) and team B (yellow). And each team contains three robots. When the program is enabled, the form of scoreboard is shown in the monitor. It is used for showing the score of each team. However, if the user doesn't want to place the scoreboard on the monitor, closing the form of scoreboard is available.

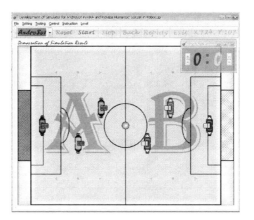

Fig. 2. The interface of the simulator, which includes the game field, the main menu and the scoreboard

3.1   Basic Settings of the Ball and the Robots

The picture of the ball and the robots are shown in Fig. 3. Fig. 4 shows the form for setting of the robots and ball. In the robot setting form, the group box of the ball is on the top of the form. In this group box, the user can adjust the radius and the RGB value of the ball.

The setting parameters of the robots are under the group box of the ball. The user can select the color of each robot. Furthermore, the length, width, and height of the robots are also can be adjusted. Next, the current position, the current angle and the initial position of the robots, are shown in this form. The user can realize the actual coordinates and the angles of the robots by opening the form.

Fig. 3. The picture of the ball and the robots

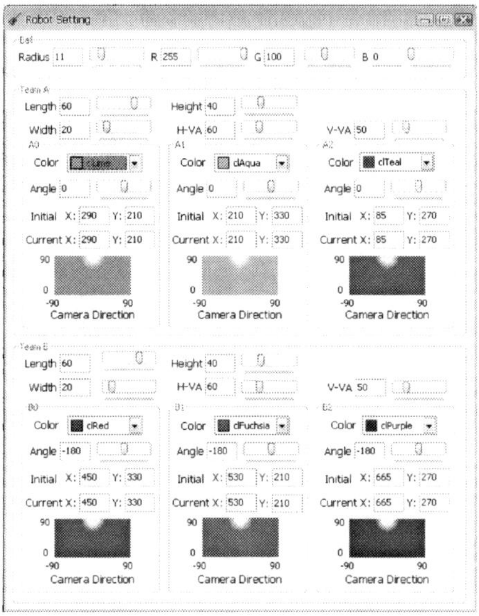

Fig. 4. The form for setting of the robots and ball

3.2 The Basic Functions

There are some basic functions in the simulator, which are described briefly below.

1)   Move the ball and robots: The user can drag the robots and the ball for changing their positions. If the user presses the robot by the mouse, rolling the wheel of the mouse can rotate the robots. If the wheel rolled up, the robot will rotate in clockwise; on the contrary, if the wheel rolled down, the robot will rotate in counterclockwise. However, making the robots overlap each other is prohibited.

2)   Turn on/off the robots: The user can double click the robots to turn on/off the robots. If the robot is turned off, it can not execute any motion.

3)   Throw the ball: The user can throw the ball during the competition. After dragging the ball and the event of mouse-up, the simulator will make the ball move by itself. Then, the speed of the ball will slow down gradually. Moreover, the moving direction and the initial speed can be controlled by the user.

4   Strategy Decision System

In AndroSot, the soccer robots are manipulated to perform the tasks of obstacle avoidance, collaboration, and competition for victory. In order to achieve the goal, a fuzzy logic based strategy is implemented for AndroSot. To lead the robot toward the target while detouring obstacle simultaneously, a potential field algorithm [3-10] of obstacle avoidance is applied. For testing the feasibility of the simulator, the strategies described above are employed in this simulator.

Fig. 5 shows the force direction at each position, which has obstacles in the field. The arrows denote the direction vector of the resultant force. We set a kicking point on the field, as shown in the yellow circle in Fig. 5. The home robot will follow the direction from its present position to the kicking point.

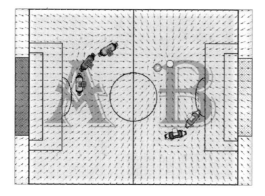

Fig. 5. Potential field navigation method

5 Extended Mode - Master Level

The collision between the robots may happen frequently in the actual competition and cause the robots damaged. In order to take this issue into consideration, the extended mode - master level is proposed. There is a line behind the robot; it denotes the HP (healthy point) of the robot.

If the collision occurred between the robots during the competition, as shown in Fig. 6, the symbol of crossing wrench will be shown on the top of the robots. Moreover, The HP of the robots will also be reduced. However, if the HP is reduced to zero, the color of the symbol will be changed into gray (as the red robot). After that, the robot cannot move anymore unless repaired.

Fig. 6. The picture of the collision between the robots

Fig. 7. The picture of the maintenance of the robots

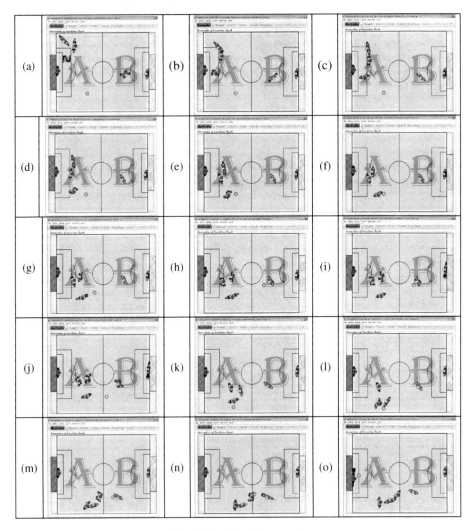

Fig. 8. The simulation result

There are four gray symbols of wrench on the corner of the field. They denote the maintenance areas. If the robot is moved into these areas, the color of the symbols will be changed into red, as shown in Fig. 7. It means that the robot is under repair. After that, the HP of the robots in the area will be supplied.

6 Simulation Result

Fig. 8 shows the simulation result. Firstly, the ball is on the left half field. Thus, the defender of Team B goes to the defending position and rotates for facing the ball. Then the attacker of Team A goes to the kicking position and shoots. However, the defender of Team B already stands on the defending position. So the ball is blocked by the defender successfully. Then, the ball is rebounded to the left half field. Therefore, the defender of Team B goes back to the defending position again. Then the attacker of Team B executes the rebound shooting. Also, the ball is blocked by the defender again. However, the ball is on the right half field. So the defender shoots, but the ball is blocked by the goalkeeper of Team A finally.

7 Conclusions

The simulator for AndroSot in FIRA is developed. the proposed simulator for AndroSot is described explicitly. Strategies which compiled to DLL files may be explicitly loaded at run-time in this simulator. In addition, another simulation mode, "Master Level" is also presented. The healthy state of each robot is considered in this mode. If the robot collided with another robots or the wall, its value of HP (healthy point) will be reduced. Finally, the effectiveness of the controlled system is shown by computer simulations, and the simulation result illustrates the feasibility of the proposed simulator.

Acknowledgment. This work was supported by National Science Council of Taiwan, R.O.C, under Grants NSC97-2221-E006-0-160-MY3 and NSC97-2221-E006-0-172-MY3.

References

1. http://www.fira.net
2. http://www.embarcadero.com/products/cbuilder
3. Rimon, E.: Exact robot navigation using artificial potential functions. IEEE Trans. on Robotics and Automation 8(5), 501–518 (1992)
4. Guo, Y.Z.: Design and implement ring potential field method for a three-on-three robot soccer game, Master Thesis, Dept. of Electrical Engineering, National Cheng Kung Univ., Tainan, Taiwan, R.O.C. (June 2002)
5. Barraquand, J., Latombe, J.C.: Robot motion planning: a distributed representation approach. International Journal Automation 10(6), 628–649 (1991)

6. Guldner, J., Utiken, V.I., Hashimoto, H., Harashima, F.: Tracking gradients of artificial potential fields with non-holonomic mobile robots. In: Proceedings of 1995 American Control Conference, pp. 2803–2804 (1995)
7. Park, M.G., Jeon, J.H., Lee, M.C.: Obstacle avoidance for mobile robots using artificial potential field approach with simulated annealing. In: Proceedings of IEEE Conf. on IEEE International Symposium, vol. 3, pp. 1530–1535 (2001)
8. Su, W., Meng, R., Yu, C.: A study on soccer robot path planning with fuzzy artificial potential field. In: Proceedings of 2010 International Conference on Computing, Control and Industrial Engineering (CCIE), vol. 1, pp. 386–390 (2010)
9. Xu, X., Xie, J., Xie, K.: Path planning and obstacle-avoidance for soccer robot based on artificial potential field and genetic algorithm. In: Proceedings of The Sixth World Congress on Intelligent Control and Automation, vol. 1, pp. 3494–3498 (2006)
10. Shi, H., Sun, C., Sun, X., Feng, T.: Chaotic potential field method and application in robot soccer game. In: Proceedings of The Sixth World Congress on Intelligent Control and Automation, pp. 9297–9301 (2006)

Design and Implementation of Big Humanoid Robot Walking Patterns Based on Inverted Pendulum Approach[*]

Kuo-Yang Tu and Ming-Fung Tsai

Institute of System Information and Control,
National Kaohsiung First University of Science and Technology,
2, Juoyue Road, Nantsu District, Kaohsiung City, Taiwan R. O. C.
tuky@ccms.nkfust.edu.tw

Abstract. The heavy weight of a big humanoid robot body makes its dynamics considerable. Therefore, the walking pattern derived according to only kinematics is difficult for a big humanoid robot walking. In this paper, the dynamics of a big humanoid robot is approached to an inverted pendulum for its walking pattern. The inverted pendulum approach regards the humanoid body as a point mass to simplify walking pattern parameters. However, parameter uncertainty is not avoided, and usually joint servo motors cannot provide enough torque for the big humanoid robot walking. In this paper, the actual servo motor move trajectories in experiments are read back to compare with the desired ones. The errors between the desire and actual trajectories are used to modify the designed trajectories. Experimental results show that the big humanoid robot can walk very well after modifying.

Keywords: Big Humanoid Robots, Walking Patterns, Inverted Pendulum, Dynamics.

1 Introduction

The leg structure makes humanoid robots easy to fall down. To maintain its stability during walking in dynamic situation requires a good mechanical design, extra sensors such as force sensors and accelerometers to acquire the information of its move. Many humanoid robots have been developed, such as ASIMO by Honda [1], WABIAN 2R by Waseda University, HUBO KHR-3 by KAIST [2] and QRIO by Sony. Hkatib proposed torque-position transformer for enough torque to steer ASIMO [3]. However, it is hard for only using joint torque to control a big humanoid robot.

The heavy body weight makes control of a big humanoid robot hard. It is because that the dynamics of the big humanoid robot influenced by gravity becomes considerable. It is possible to produce considerable influence owing to a little bit of

---

[*] This research was supported by National Science Council, Taiwan, Rep. of China under grant NSC 98-2622-E-018-CC2-.

T.-H.S. Li et al. (Eds.): FIRA 2011, CCIS 212, pp. 241–249, 2011.
© Springer-Verlag Berlin Heidelberg 2011

kinematic error in a big humanoid. However, it is hard to have precise kinematics for walking pattern design. Humanoid robots are open-chain structure that makes their postures hard to be estimated by joint angles. Even if using successful algorithms estimate humanoid robot posture, its error is usually huge by accumulative inherence. Hence, accelerometers and gyroscopes are usually designed to estimate the absolute posture of a humanoid robot. The other function of accelerometers and gyroscopes are to measure the walking dynamics of humanoid robots. Especially, the heave dynamics of a big humanoid robot is hard to control in open loop.

In intuition, ZMP (Zero Moment Point) provides a solution to support humanoid body during walking [4, 5]. However, ZMP which only takes geometric relationship is hard to solve the heavy dynamics of a big humanoid robot. Thus, the method to measure ZMP is proposed [6]. Kim et. al. proposed the humanoid walking pattern to imitate human walking [7]. On the other hand, optimization of wasting energy is also an important issue for humanoid walking pattern [8]. However, the approach to humanoid walking dynamics is direct methods [9 - 11]. In this paper, the humanoid walking pattern based on inverted pendulum is proposed. The main advantage of this approach is to parameterize the heavy dynamics of the big humanoid robot body during walking. Then the trajectory that the humanoid robot can walk continuously is derived. Finally the derived walking pattern is implemented by actual experiments.

The rest of this paper is organized as follows. The fundamental mathematics of the big humanoid robot is constructed in the next section. The walking pattern of the big humanoid robot is designed and implemented in section 3 and 4, respectively. In section 5, the walking pattern is examined to an exact big humanoid robot by experiments. The experiments to modify trajectory for better walking are also included. Finally, the conclusions and further development is presented in section 6.

2   Kinematics Model of a Big Humanoid Robot

In this section, how to design a big humanoid robot is described. The kinematics of the designed humanoid robot is derived. In addition, the kinematic equation is verified by joint trajectory calculation for a walking step.

The mechanical design considers the minimum Degree Of Freedom (DOF) to let a humanoid robot have smooth walking. Therefore, one leg and one hand are consisted of 6 DOFs and 3 DOFs, respectively. In addition, there are two DOFs designed for its head so that the camera can move to have wide view angle.

The kinematics is important for walking pattern planning and controller design. The following thus derives the kinematic equation of the designed humanoid robot. In traditional method, humanoid robots usually derive forward kinematic equations in geometric space. Although the solution of using geometric space is comprehensive, vector method that is easy to derive robot dynamics is engaged in this paper.

Let the right foot locate at $\vec{p}_0 = [x_0\ y_0\ z_0]$. Then the locations of every join of the humanoid robot are

$$\vec{p}_0 = [x_0\ y_0\ z_0]$$

$$\vec{P}_1 = \vec{P}_0 + R_1$$

$$\vec{P}_2 = \vec{P}_1 + R_{x,\theta_1} R_{y,\theta_2} R_2$$

$$\vec{P}_3 = \vec{P}_2 + R_{x,\theta_1} R_{y,\theta_2} R_{y,\theta_3} R_3$$

$$\vec{P}_4 = \vec{P}_3 + R_{x,\theta_1} R_{y,\theta_2} R_{y,\theta_3} R_{y,\theta_4} R_{x,\theta_5} R_4$$

$$\vec{P}_h = \vec{P}_4 + R_{x,\theta_1} R_{y,\theta_2} R_{y,\theta_3} R_{y,\theta_4} R_{x,\theta_5} R_{z,\theta_6} D_R$$

$$\vec{H} = \vec{P}_h + R_{x,\theta_1} R_{y,\theta_2} R_{y,\theta_3} R_{y,\theta_4} R_{x,\theta_5} R_{z,\theta_6} D_L$$

$$\vec{P}_5 = \vec{H} + R_{x,\theta_1} R_{y,\theta_2} R_{y,\theta_3} R_{y,\theta_4} R_{x,\theta_5} R_{z,\theta_6} R_{z,\theta_7} R_5$$

$$\vec{P}_6 = \vec{P}_5 + R_{x,\theta_1} R_{y,\theta_2} R_{y,\theta_3} R_{y,\theta_4} R_{x,\theta_5} R_{z,\theta_6} R_{z,\theta_7} R_{x,\theta_8} R_{y,\theta_9} R_6$$

$$\vec{P}_7 = \vec{P}_6 + R_{x,\theta_1} R_{y,\theta_2} R_{y,\theta_3} R_{y,\theta_4} R_{x,\theta_5} R_{z,\theta_6} R_{z,\theta_7} R_{x,\theta_8} R_{y,\theta_9} R_{y,\theta_{10}} R_7$$

$$\vec{P}_8 = \vec{P}_7 + R_{x,\theta_1} R_{y,\theta_2} R_{y,\theta_3} R_{y,\theta_4} R_{x,\theta_5} R_{z,\theta_6} R_{z,\theta_7} R_{x,\theta_8} R_{y,\theta_9} R_{y,\theta_{10}} R_{y,\theta_{11}} R_{x,\theta_{12}} R_8 \tag{1}$$

where R_i (for $i = 1, \ldots, 8$) are the links of the humanoid robot.

In order to demonstrate the derive kinematic equations, we draw the humanoid robot in initial pose by using MATLAB. Fig. 1 shows the result of the humanoid robot at an initial pose in actual mechanical dimensions. This result demonstrates that the kinematic equation is correct.

The forward kinematic equation is used to derive inverse kinematics of the humanoid robot. And then the derived inverse kinematics is demonstrated by calculating the humanoid robot for a walking step. Fig. 1 shows the stick diagram of the humanoid robot in a walking step. Figs. 1 demonstrate the correction of both forward and inverse kinematics of the humanoid robot.

After designing, the humanoid robot is implemented. Its height is 1.3 meters, and weight is 6.8 Kg. The mechanical design of the big humanoid robot is based on a golden ratio rule that is usually used to evaluate fashion models. The rule of the ratio between upper body and total tall is 0.618. Hopefully, the humanoid robot designed based on this rule can be like a fashion model girl that always owns elegant postures.

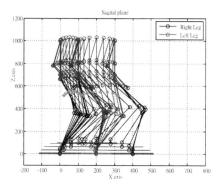

Fig. 1. The stick diagram of the humanoid robot for a walking step

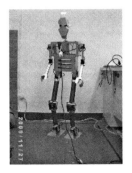

Fig. 2. The photo of designed humanoid robot

The big humanoid robot as shown in Fig. 2 is totally made in Intelligent Robotics Laboratory at National Kaohsiung First University of Science and Technology (NKFUST). All of the mechanical components of the big humanoid robot are produced by a mini-CNC machine and a bending machine in our laboratory. After the component shapes cut by the mini-CNC machine, the bending machine bends them as the structure components for the humanoid robot. In addition, the head includes two COM cameras for 3-D image processing in the future.

3 Walking Pattern Design

In this paper, the walking pattern design makes use of inverted pendulum to approach to the big humanoid robot body dynamics. This approach engages [9-10], but modifies some parameters for continuous trajectory.

The heavy humanoid dynamics is approached to an inverted pendulum model. Under this approach, there are three assumptions: the whole body mass approached to the point at the center of gravity, ignoring the leg weight and frontal plane movement. In sagitte plane (x-z plane), the humanoid walking can be approached to Fig. 3. As show in Fig. 3, the body dynamics can be expressed by

$$M\ddot{x} = f \sin \theta \tag{2}$$

$$f \cos \theta - Mg = 0 \tag{3}$$

where M is the mass of humanoid body, f is the force on the body, and θ is the angle between support and horizontal line (z). Eqs. (2) and (3) can be merged and manipulated to become the following

$$\ddot{x} = g \tan \theta = g \frac{x_0}{z_0} = g \frac{x_0}{Zc} \tag{4}$$

Where x_0 and z_0 is the position of the center of gravity, and Z_C is a designed constant to maintain the body on same height during walking.

Integrating Eq. (4) one and two times can obtain the velocity and position trajectories, respectively, as follows:

$$\dot{x}(t) = x(0)/Tc\sinh(t/Tc) + \dot{x}(0)\cosh(t/Tc) \tag{5}$$

$$x(t) = x(0)\cosh(t/Tc) + Tc\dot{x}(0)\sinh(t/Tc) \tag{6}$$

where $T_C = \sqrt{Z_c/g}$ is a constant, x(0) is initial position, $\dot{x}(0)$ is initial velocity, and t is time variable.

Moreover, the energy variation is considerable factor during humanoid walking. Multiplying Eq. (4) by $\dot{X}$ can obtain

$$\frac{1}{2}\dot{x}^2 - \frac{g}{2Z_C}x^2 = \text{constant} \tag{7}$$

In Eq. (7), the first term is kinetics energy, and the second term is potential energy. This equation implies that the move of inverted pendulum must reserve constant even the kinetics and potential energy changes. Therefore, using inverted pendulum approach to design the body move trajectory during humanoid walking must satisfies that the sum of kinetics and potential energy is constant. When the body height is 0.7 m, the sum of kinetics and potential energy is constant at the initial conditions x(0) = -0.1 and $\dot{x}(0)$ = 0.4. Fig. 4 shows this result. In this paper, this result will be extended 3 dimensions to design the big humanoid walking pattern.

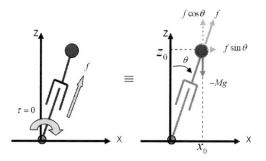

Fig. 3. The humanoid robot walking in front plane

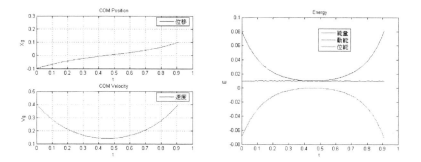

Fig. 4. The position and velocity trajectories and energy variation during humanoid walking at x(0) = -0.1, $\dot{x}(0)$ = 0.4 as Z_c = 0.7 m

The derived trajectory as above is the waist trajectory for humanoid waling. In addition to the waist trajectory, the humanoid walking needs a sole trajectory. In this paper, the sole trajectory employs cycloid equation. The cycloid equation for the humanoid walking expresses one step walking as follows (for $0 \le t \le t_1$):

$$X = (k-1)a_l + \frac{a_l}{\pi}\left(\left(\frac{2\pi t}{T}\right) - \sin\left(\frac{2\pi t}{T}\right)\right) \tag{8}$$

$$Z = \frac{h}{2}\left(1 - \cos\theta\left(\frac{2\pi t}{T}\right)\right) + h_1 \tag{9}$$

where a_l is one half of a walking stride, h is the height of the walking stride, and h_1 is the distance between ankle and the foot bottom.

The benefit of cycloid trajectory is uniform speed, i. e. constant acceleration. Theoretically, the uniform speed lets the humanoid robot don't have sliding force during waling. This result can be examined by calculating the double differential of x(t) and z(t) in Eqs. (8) and (9).

4   Walking Pattern Implementation

Any humanoid robot needs walking patterns for walking. In this paper, the big humanoid robot walking pattern is designed by two trajectories: the waist and sole trajectories. In this section, how to implement the design walking pattern is described.

Fig. 5 shows the walking pattern designed for humanoid walking. As shown in Fig. 5, the black doted line is sole trajectory, and the red doted line is the waist trajectory. The waist trajectory is designed to maneuver humanoid body continuously, but the sole trajectory move the swing foot at the next support place.

Two trajectories form the walking pattern of the humanoid robot. After solving the inverse kinematics from forward kinematics Eqs. (1), two trajectories can used to solve the trajectories of all joints of the humanoid robot.

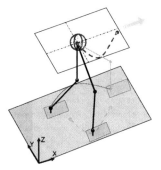

Fig. 5. The cycloid trajectory used for the humanoid walking pattern

5   Experimental Results

All joint trajectories solved by waist and sole trajectories are engaged to control the humanoid robot servo motor for walking. However, the heavy dynamics of the big humanoid robot cannot be overcome easily. In this section, the experimental results include how to compensate trajectory error for stable walking.

In the first time experiment, the humanoid robot cannot walk very well. After reading back the actual move trajectories, we found that the servo motor cannot be controlled to follow the desired trajectories. There exist steady state errors. Fig. 6 shows the hip trajectories of right and left legs in X-Z plane. As shown in Fig. 6, the blue circles indicate the duration of large errors.

The steady state errors result in that the swing leg cannot immediate to leave ground. Then the swing leg touching ground produces disturbance to the control of all joint. Fig. 6 depicts the swing leg cannot leave ground because of hip joint trajectory errors. Therefore, the desired trajectories are modified to compensate the steady errors of humanoid robot servo motors. The modification increases or decrease with regards to the joint trajectory errors. For example, hip joint trajectories of the left and right legs in X-Z plane increase and decrease, respectively, because their errors are plus and minus, respectively, as shown in Fig. 6. After modifying, the humanoid robot walks very well, and the desired and actual trajectories are very close as shown in Fig. 7.

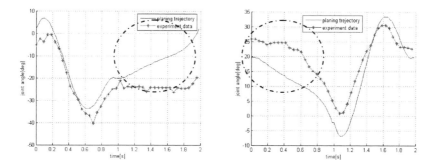

Fig. 6. The original trajectories of hip in X-Z plane (left: left leg; right: right leg)

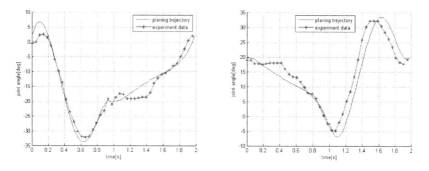

Fig. 7. The hip joint trajectories in X-Z plane after compensating

6 Conclusions and Further Development

In this paper, the walking pattern based on inverted pendulum to parameterize the heavy dynamics of a big humanoid robot body is derived. Actually, the walking pattern is designed by two trajectories for waist and sole, respectively. In addition to deriving the humanoid body walking based on inverted pendulum for its waist trajectory, the cycloid equation is designed for its sole trajectory so that the swing leg moves in constant velocity. After design and implementation, the experiments examine the walking pattern in an actual big humanoid robot. Because of steady state errors of humanoid joint servo motors, the big humanoid robot cannot be controlled to follow the desired trajectories. Thus, some compensation is added to the desired trajectory according to the static state errors read back from the servo motors. Finally, the humanoid robot can be controlled to walk.

However, the torque of joint servo motors is not big enough to steer the humanoid for walking. Thus changing bigger servo motors is necessary for better walking in the next generation of big humanoid robot. In addition, the parameters of big humanoid heavy dynamics cannot be figured out easily. However, the heavy dynamics of a big humanoid robot can be read by accelerometers and gyroscopes. Therefore, in the future, a motion sensor module consisted of accelerometers and gyroscopes is designed to read the humanoid dynamics for stable walking control.

Acknowledgments. This research was supported by National Science Council, Taiwan, Rep. of China under grant NSC 98-2622-E-327-018-CC2.

References

1. Honda homepage, http://www.honda.co.jp/ASIMO
2. http://mind.kaist.ac.kr/3re/research.htm
3. Hkatib, O., Thaulad, P., Yoshikawa, T., Park, J.: Torque-Position Transformer for Task Control of Position Conlled Robots. In: 2008 IEEE Intern. Confer. on Robotics and Automation, pp. 1729–1734 (2008)
4. Tak, A., Tochizawa, M., Karaki, H., Kato, I.: Dynamic Biped Walking Stabilized with Optimal Trunk and Waist Motion. In: IEEE/RSJ International Workshop on Intelligent Robots and Systems, pp. 187–192 (1989)
5. Huang, Q., Yokoi, K., Kajita, S., Kaneko, K., Arai, H., Koyachi, N., Tanie, K.: Planning Walking Pattern for a Biped Robot. IEEE Trans. on Robotics and Automation 17(3), 280–289 (2001)
6. Erbatur, K., Okazaki, A., Obiya, K., Takahashi, T., Kawamura, A.: A Study on the Zero Moment Point Measurement for Biped Walking Robots. In: IEEE International Workshop on Advanced Motion Control, pp. 431–436 (2002)
7. Kim, J.G., Noh, K.-G., Part, K.: Human-like Dynamic Walking for Biped Robot Using Genetic Algorithm. In: Liu, Y., Tanaka, K., Iwata, M., Higuchi, T., Yasunaga, M. (eds.) ICES 2001. LNCS, vol. 2210, p. 159. Springer, Heidelberg (2001)
8. Ha, S.-S., Yu, J.-H., Han, Y.-J., Hahn, H.-S.: Natural Gait Generation of Biped Robot Based on Analysis of Human's Gait. In: International Conference on Smart Manufacturing Application, pp. 30–34 (2008)

9. Kajita, S., Tani, K.: Experimental Study of Biped Dynamic Walking. IEEE Control System Technology, 13–19 (1996)
10. Kajita, S., Tani, K.: Study of Dynamic Biped Locomotion on Rugged Terain-Derivation and Application of the Linear Inverted Pendulum Mode. In: IEEE Conference on Robotics and Automation, vol. 3, pp. 2885–2891 (1995)
11. Noh, K.-K., Kim, J.-G., Huh, U.-Y.: Stability Experiment of a Biped Walking Robot with Inverted Pendulum. In: The 30th Annual Conference of the IEEE Industrial Electronics Society, vol. 3, pp. 2475–2479 (2004)

Optimum Iris Opening for Soccer Robot Detection under Un-uniform Lighting

Abdul Rahim Bin Ibrahim[1], Choong-Yeun Liong[2],
and Khairun Syatirin Bin Md Salleh[1]

[1] Mechanical Engineering Department, Port Dickson Polytechnic
KM 14, Jalan Pantai, 71050 Si Rusa, Negeri Sembilan, Malaysia
[2] Centre for Modelling and Data Analysis (DELTA), School of Mathematical Sciences,
Faculty of Science and Technology,
Centre for Artificial Intelligence Technology (CAIT),
Faculty of Information Science and Technology
Universiti Kebangsaan Malaysia, 43600 UKM Bangi, Selangor DE, Malaysia
{abdulrahim,khairun}@polipd.edu.my, lg@ukm.my

Abstract. With the rapid technological advancement in vision systems, the advantages of vision based system as the sensing device have influenced the quality of the Robot Soccer game. This study presents an approach to produce a balanced image in terms of brightness under un-uniform lighting environment through the control of lens iris opening. Existing image processing module and CCD camera were used to obtain data of HSL (color system) properties in three un-uniform lighting conditions. These data were used to deduce the percentage of optimum iris lens aperture to increase efficiency of soccer robot detection under the un-uniform lighting conditions. The optimum range of iris lens opening was identified and the robots were detected successfully for all the un-uniform lighting conditions.

Keywords: Lens iris opening, HSL color system, un-uniform lighting.

1 Introduction

Robot Soccer is a game that combines several elements: vision systems, artificial intelligence for decision making and problem solving, communication system, real time motion control and mechanical elements (Figure 1). The purpose of the consolidation is to create an autonomous robot system that can play soccer. In Robot Soccer game, the robot players must work together in order to find the ball, their own goal as well as to create goals in the opponent side [1].

Vision system is among the most important element in the autonomous robot system because only the CCD camera is used as a sensing device. This means that the input data, such as robots position and orientation, coordinate, robots obstacle and robots ID detection tied closely to the images taken by the CCD camera.

The quality of the images captured by a CCD camera system is closely related to the playing field lighting. Poorly designed lighting can result in unintelligible images, with under and over illuminated regions showing poor contrast, and failed to get the

T.-H.S. Li et al. (Eds.): FIRA 2011, CCIS 212, pp. 250–257, 2011.

image information to be detected effectively [2]. In fact, lighting of soccer robot playground often faces the problem of un-uniform lighting (Figure 2). Un-uniform lighting problems exist because of irregular illumination and it is a major problem in vision-based autonomous robot system. Irregular lighting is the main reason for incorrect estimation for the robot (system) to detect the other robots and the obstacles.

Many studies have been made previously to determine the appropriate approach to solve the problem of un-uniform lighting. Among these approaches are lighting configuration, least squares method, face location system, Discrete Cosine Transform, Local Binary Pattern, shape and near-blue screen lighting technique [3].

Another approach that can be used to solve the problem of un-uniform lighting is lens iris control methods. It is done by controlling the lens iris opening to reduce the difference in brightness between under and over illuminated regions. Generally, the lens typically has three adjustable parameters: zoom, focus and iris (Figure 3). In this study, zoom and focus have become fixed variable that were set to produce a sharp visual image. Visual image is generally defined as the reproduction of two dimensional visual information on the screen [4]. In this study, the images refer to visual images that are meant to be able to give information such as the robots position and orientation, coordinate and robots obstacle.

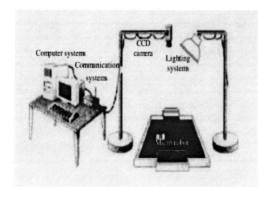

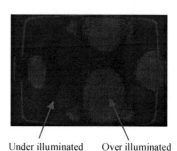

Under illuminated   Over illuminated

Fig. 1. Robot Soccer System Overview

Fig. 2. Un-uniform lighting on playing surface

Fig. 3. CCD Camera Lens

The image processing module used in this study is SP Mirosot RSS vision server that is only compatible with Basler CCD camera. Besides that, we work in the HSL color system as the image processing module available for color tuning works only for the HSL color system. Quick and effective image processing that comes with accuracy and robustness are keys to meet the requirement of the robot soccer system in real time. An important issue in real time environment is the uniformity of the lighting condition. However, such condition is often difficult to achieve, especially in outdoor environment. Therefore, the main aim of this paper is to show the specific lens iris opening for effective detection of soccer robot in HSL color space under un-uniform lighting conditions. A simple range of values for the effective lens iris opening, that can be set up easily, would be a great help to the users who are not trained in visual perception or in lighting design.

2 HSL Color System

HSL is one of the models for describing the color space where H stands for hue, S for saturation, and L for luminosity or luminance. The HSL color space forms a double cone that is neutral in the middle, black at the bottom and white at the top [5].

Roughly, hue corresponds to the base color, ranging from 0 to 360 degrees as shown in Figure 4. Saturation corresponds to the intensity of the color, ranging from 0% to 100%, where 0% means gray and 100% means saturated color; whereas luminosity corresponds to the brightness of the color, ranging from 0% to 100%, where 0% is the extreme darkness (black) and 100% the extreme lightness (white).

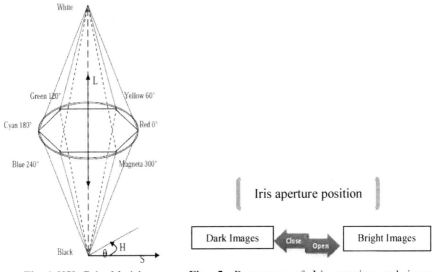

Fig. 4. HSL Color Model

Fig. 5. Percentage of Iris opening and image brightness

3 Lens Iris

Manual Iris Lens is the simplest type of iris control lens with manual adjustment to set the iris opening in a fixed position. These are generally used for fixed lighting applications or where the camera to be used is readily accessible, and lighting level stays mostly constant. Iris is an adjustable aperture used to control the amount of light coming through the lens. The more the iris is opened, the more light it lets in and the brighter the scene will be. Manual iris adjustment can be demonstrated as in Figure 5.

4   Un-uniform Lighting

In real-world environment, producing uniform lighting at an area of the surface is very difficult. This is because there are external factors that are difficult to control that will disturb the equilibrium (uniformity) of light. Hence, under an uncontrolled environment, lighting variability and un-uniformity is common [6].

The surface used in this study is 1.8m × 2.2m in size and the surface color is black. This surface is actually a robot soccer playing field. Un-uniform lighting conditions on this surface will affect the overall color detection generally and the soccer robot detection particularly. If the issue of un-uniform lighting becomes critical, the vision system will not be able to detect the position and orientation of the soccer robot as it fails to detect the color (the robot). This means that the robot soccer system will function in the blind. Therefore, it can be summarized that the recognition and tracking of soccer robot in a balanced lighting condition is very important in order to perform the operation of the autonomous robot soccer system.

5   Experimental Method

The experimental study was carried out as depicted in Figure 6. A total of three un-uniform lighting conditions of similar intensity as shown in Figure 7 were used.

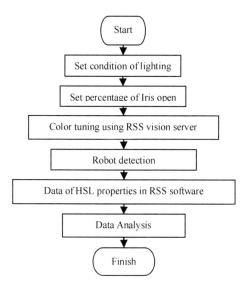

Fig. 6. Flowchart of the Experimental Method

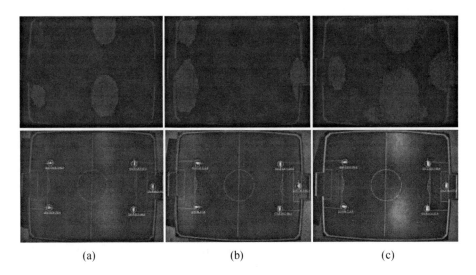

(a)   (b)   (c)

Fig. 7. Experimental lighting conditions and robot detections for (a) Lighting condition 1, (b) Lighting condition 2, and (c) Lighting condition 3

For each lighting condition, the color tuning was done using the SP software of the Mirosot RSS vision server according to the percentage of iris opening in the range of 10% - 90% (interval 10%). Iris opening 0% and 100% were not counted because they produce images that are either too dark (0%), or too bright (100%).

The purpose of color tuning is to produce the perfect robot detection. Finally, the data of HSL (Hue, Saturation and Luminance) properties were acquired for the purpose of analysis. As the data are in the form of HSL properties range, all properties are coded as values of Hmin, Hmax, Smin, Smax, Lmin and Lmax.

6   Experimental Results and Analysis

Results of HSL properties from experiments done using three different lighting conditions are shown in Table 1. The findings show that the HSL luminance properties have a very significant linear relationship ($r > 0.9$) with lens iris opening as shown in Table 2. On the other hand, Hue and Saturation properties exhibit less significant relationship to lens iris opening compared to the luminance value, and hence were ignored. Figure 8 shows the changes in image luminance relative to level (percentage) of lens iris opening. It is clear that percentage of lens iris opening has an almost linear relationship with image luminance, especially in the range between 10 to 50% of iris opening.

If examined closely, luminance (L) were almost similar when lens iris opening were between 10% - 30% for the three different lighting conditions. This means that the brightness of the color detected in that range of iris opening has not changed much. However, when lens iris opening was 10%, robot detection was not perfect. This is because the image was very dark. From the experiments conducted, the robots were detected perfectly when the iris opening were between 20% - 30% for all the three lighting conditions.

Table 1. Experimental data of HSL properties for (a) Lighting condition 1, (b) Lighting condition 2, and (c) Lighting condition 3.

Lighting Condition	Lighting Condition 1 (LC1)								
Iris Open (%)	10	20	30	40	50	60	70	80	90
Hue Min LC1	-106	56	-42	-42	16	23	21	24	24
Hue Max LC1	0	42	-53	-84	84	70	84	68	66
Saturation Min LC1	0	0	0	0	3	11	6	10	13
Saturation Max LC1	54	38	39	43	47	48	53	49	51
Luminance Min LC1	15	18	23	25	35	40	35	43	44
Luminance Max LC1	20	26	39	53	64	74	80	84	83
Lighting Condition	Lighting Condition 2 (LC2)								
Hue Min LC2	-106	21	53	42	0	42	0	53	54
Hue Max LC2	-42	0	42	0	-42	-84	-42	-106	-110
Saturation Min LC2	6	0	0	0	0	3	0	2	5
Saturation Max LC2	60	48	36	36	31	38	35	38	38
Luminance Min LC2	15	18	22	26	30	35	30	41	41
Luminance Max LC2	20	26	34	43	53	66	79	72	71
Lighting Condition	Lighting Condition 3 (LC3)								
Hue Min LC3	-127	99	10	0	0	0	0	0	0
Hue Max LC3	0	84	0	-42	-42	-84	-84	-84	-42
Saturation Min LC3	0	0	0	0	0	0	0	0	0
Saturation Max LC3	63	48	37	37	42	41	46	47	46
Luminance Min LC3	14	18	22	25	29	29	32	31	30
Luminance Max LC3	22	28	41	56	77	82	103	102	102

Table 2. Pearson Correlation between lens iris opening and HSL properties

HSL Properties	Pearson Correlation	Strength
Lmin avg	0.9721	Very Strong
Lmax avg	0.9719	
Smin avg	0.7503	Moderate
Smax avg	-0.2723	
Hmin avg	0.4609	Moderate
Hmax avg	-0.5540	

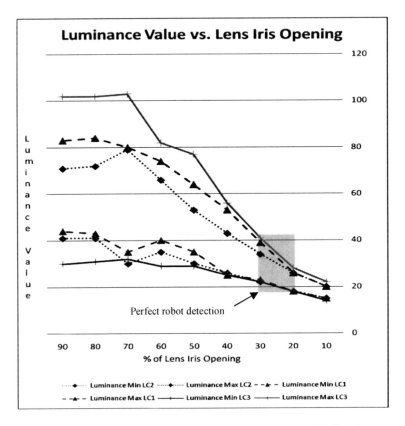

Fig. 8. Graph shows relation between Luminance and Lens Iris Opening

7 Conclusion

In this experimental study, an iris lens opening control approach has been proposed to control the difference in image brightness between over and under illuminated regions. Although in principle iris lens opening is known to affect image brightness, our experiments show that the optimum 20% - 30% lens iris opening has dispelled the effects of differences in image brightness between the over and under illuminated regions, and produced perfect soccer robot detection capabilities. For the purpose of further studies, un-uniform lighting conditions under different lighting intensities and design should be conducted to describe the relation between the lens iris opening and the lighting environment. In addition, motorized lens iris can also be used to control the lens iris opening to facilitate further research.

Acknowledgment. We would like to thank the members of the Pattern Recognition Group, Universiti Kebangsaan Malaysia for their constructive comments, encouragement and support, and the Ministry of Higher Education, Malaysia for the financial support.

References

1. Kim, J.-H., Shim, H.-S., Kim, H.-S., Jung, M.-J., Choi, I.-H., Kim, J.-0.: A Cooperative Multi-Agent System and Real Time Application to Robot Soccer. In: Proc. of the 1997 IEEE Int. Conf. on Robotics and Automation, Albuquerque, New Mexico, pp. 638–643 (1997)
2. Shacked, R.: Automatic Lighting Design Using a Perceptual Quality Metric, M.Sc. thesis, School of Computer Science and Engineering. Hebrew University of Jerusalem, Israel (2001)
3. Pratomo, A.H., Zakaria, M.S., Prabuwono, A.S., Omar, K.: Illumination Systems For Autonomous Robot: Implementation and Design. Journal of Engineering and Applied Sciences 4, 342–347 (2009)
4. Peng, S., Yongge, W.: Research Based on the HSV Humanoid Robot Soccer Image Processing. In: Proc. of 2010 Second International Conference on Communication Systems, Networks and Applications, pp. 52–55 (2010)
5. The Chaos Pro Website (2011),
   `http://www.chaospro.de/documentation/html/paletteeditor/colors`
   `pace_hsl.htm`
6. Harahap, D.A., Prabuwono, A.S., Abdullah, A.: Illumination Normalization Methods for Object Recognition in Robot Soccer Vision. In: Proc. of International Conference on Pattern Analysis and Intelligence Robotics (submitted, 2011)

Framework for Measuring Effectiveness of Preventive Maintenance for Velocity of Robot Soccer

Mohd Zaidi Bin Mahmud[1], Che Seman Bin Che Cob[1],
Mohd Faizul Emizal Bin Mohd Ghazi[1], and Shahnorbanun Sahran[2]

[1] Mechanical Engineering Department
Politeknik Kota Bharu
16450 Kota Bharu, Kelantan
{mzaidi,cheseman,faizul}@pkb.edu.my
[2] Pattern Recognition Research Group, CAIT,
Faculty of Information Science and Technology,
universiti Kebangsaan Malaysia,
43600 Bangi, Selangor
shah@ftsm.ukm.my

Abstract. Soccer Robotics is the field that combines multiple important discipline including artificial intelligence and mobile robotics. The framework for the complementary purposes of research and education in multi-agent robotics arises for exploration of new technology. Preventive maintenance is essential thing to preserve aging of Robot Soccer Mirosot and also to decrease the costing to maintain performance of the robot. This paper will be discussing a framework classified as model-specific to execute preventive maintenance for Robot Soccer as follow the standard of procedure. A plan schedule of preventive maintenance established quarterly will be considered the physical robot situated in a real environment and it's reactive, proactive and communicative. The primary goal of this framework preventive maintenance it will be use as standard of procedure to anyone during handling maintenance of Robot Soccer Mirosot. It also to prevent the failure of Robot Soccer before it actually occurs. It is designed to preserve and enhance equipment reliability by replacing defect components before they actually fail.

1 Introduction

Robot Soccer is a multi-agent system that consists of artificial intelligence and mobile robotics [1]. In preserve this item is crucial matters must be consider as procedure in maintaining theirs performance. The velocity of the robot will be experimented as a crucial thing in established the frameworks of the preventive maintenance. A variety of mechanical parts like wheel, body, gearing system, rod of roller, bearing, terminal ports, PCB board, DC battery and others will through a systematic inspection, detection and correction process to determine the failures or damages. The features of preventive maintenance will include testing, measurements by specific tools, adjustments, and parts replacement performed technical personnel follow a standard of procedure.

T.-H.S. Li et al. (Eds.): FIRA 2011, CCIS 212, pp. 258–269, 2011.

There are three types of preventive maintenance including model-specific, annual and three-year inspection and environmental concerns. This paper focusing on model-specific as a subject to study. A clean robot and well-oiled gearing system will affect the velocity of robot to enhance their performance. All the procedure of preventive maintenance to lubricate every axis, jointer and the rolling bearing is simple duty but essentially to preserve robot from dusty, any corrosion and grime from decreasing their performance. Any visible damage will be inspecting especially electronic parts because it's a major part for movement of robot and inspection the components, connector cable and terminal ports for any damage or losses.

There is some issue or problem occurred before executing the preventive maintenance to our robot soccer system. The failure of movement during dribbling and chase the ball accurately is the main problem. The primary goal to our issue is define the solution for this problem in a proper   and long term activities in order to preserve the robot condition. The other issue is increasing of the number of defected parts caused by collision among the robot itself. The most important issue of the robot soccer is to maintain the optimum velocity of the robot during dribbling and chasing the ball. Stuck and jam gearing system need to lubricate with lubrication oil to ovoid decreeing of the robot velocity.

Planning of preventive maintenance of the robot soccer is a crucial task. A standard framework for preventive maintenance will be proposed to avoid the occurrence of failure of the robot soccer and reduce potential consequences of failures. We focusing on developing model specific in order executing a proper preventive maintenance.

2   Literature Review

Several methods have been proposed in the literature for executing preventive maintenance for industrial robotic. Chin-Tai Chen proposed preventive maintenance using Markov chain as multi-state as numeric solution to predict the deteriorating and probability to damage of system dynamic method [2]. The numerical method assumes a prediction of failure and deterioration of robotic system in minimizing the costing and preserves the aging of system. Salman T. Al-Mishari and S.M.A Suliman proposed the model that used for PM through load sharing concept or parallel load based on costing and benefit to the components in avoiding damage occurred at auxiliary components [3]. O.Roux, D.Duvivier, G. Quesnal, E.Ramat aims to provide a framework to optimization of production and maintenance through simulation [4]. The GUI concept come with Nelder-Mead method are the main characteristic in optimization of preventive maintenance through simulation model.

ImanNasoohi, Syed Reza Hejazimentioned in their journal that naturally maintenance classified into two categories, corrective and preventive [5]. Corrective maintenance is performed after system failure. Preventive maintenance is activities before systems failure. Preventive maintenance is including inspection and replacement. Usually the preventive maintenance effectiveness should be evaluated from different perspectives. Michael Bartholomew-Biggs, Ming J.Zuo, Xiaohu Li approach presented here involves a performance function [6]. In practice, preventive maintenance is used to lengthen the equipment useful lifetime. This paper describes a numerical investigation of velocity to maintain the original performance all the time. J.T.Selvik, T.Avenstate in their journal the preventive maintenance is introduced to avoid the failure of the system, reduce potential consequences of failures, and reduce maintenance costs, also increase reliability and safety [7].

Mann, Saxena and Knapp describe two approaches have evolved for performing preventive maintenance. The traditional approach is based on the use of statistical and reliability analysis of equipment failure [8]. Under statistical-reliability (S-R)-based preventive maintenance, the robot soccer component will be replace based on statistic data collected from component inspection. In this paper we present a model-specific for preventive maintenance base on other model found in literature for robot soccer as presented in Figure 1.Using this framework we can monitor robot soccer performance and also can be an effective tool for detecting component faults. A fault component is recognized when certain limit value of robot velocity are acceded after doing the inspection. The velocity of the robot soccer can be used for measuring the effectiveness of this framework. Once inspections have been done and data are being collected, it is necessary to have a reliable mean of interpreting the data to detect when fault condition is occurring.

Table 1. Literature review summary

#	Title	Technic	Result	Comment
1	Dynamic preventive maintenance strategy for an aging and determination production system (2011)	Propose a dynamic preventive maintenance by using Markov chain formula to predict aging and detracting of production system	Can estimate the current probability transition matrix and the aging factor based on historical data.	This method only focusing on mathematical theory base on collection of historical data.
2	A framework for reliability and risk centered maintenance (2011)	Reliability centered maintenance is an analysis method for PM	By using this method reliability and consequence of relevant system item are assessed to identify and determine suitable scheduling PM task and interval	An approach of RCM method suitable implemented in industrial machinery and huge system.
3	A multi-objective approach to simultaneous determination of	By using ε-constraint method to generate different pareto-	This model can determines number of spare parts and	There is no framework or standard procedure

	spare part numbers and preventive replacement times (2011)	optimal solution can solve multi-objective model of PM	replacement of defected parts	provided in this paper.
4	Optimization of preventive maintenance through a combined maintenance-production simulation model (2010)	Combination of timed petri-nets and PDEVS model to optimize preventive maintenance via Nelder-Mead (Simplex) method	Optimization of PM by using this technic can decrease failure to the system	This method only tested base on VLE simulator. (Not implemented on real environment)
5	Modelling and preventive for auxiliary component (2009)	The concept of load sharing to build a Markov model by using regression analysis to illustrate the model	Useful for design of and optimum preventive maintenance model	This method is effective for implementing PM for auxiliary component
6	Modelling and optimizing sequential imperfect preventive maintenance (2009)	By using Kajima method that combine numerical solution and statistical analysis to identify imperfectly scheduling preventive maintenance	This method does not just reduce failure but also make equipment operate more efficiently.	A complex numerical equation to provide a PM schedule and need a long term periodic analysis
7	Preventive maintenance scheduling for multi-cogeneration plants with production constraints (2007)	This mathematical method base on a mixed integer programming model used for cogeneration plants to verify scheduling implemented at industrial plants [9]	Produce an efficiently scheduling of PM at cogeneration plants.	Suitable for implementing in industrial plant and huge system
8	Statistical-based or condition-based preventive maintenance? (1995)	The traditional approach based on the use of statistical and reliability (S-R) analysis of equipment failure	This method contribute a decreasing percentage to total maintenance needs	S-R-based preventive maintenance is only economical where the standard deviation of the failure population is small.

From the literature review and our experience, there is no research related to executing preventive maintenance to the robot soccer before. The previous preventive maintenance tends to used numerical or mathematical solution implemented in industrial equipment. There are also researches modeled by using simulation to predict the failure of the system. Because of this factor we are motivated to develop a framework of preventive maintenance and measuring the effectiveness through velocity experimental. We hope this paper will contribute a huge contribution on preventive maintenance of robot soccer.

3   Methodology of Research

Fig.1 had shown the flow chart of methodology in implement preventive maintenance based on standard procedure. The visible check of condition of outlet parts has been done to ensure the parts or components in good condition. After that, robots have been tested with velocity experimental to measure current performance before executing the preventive maintenance. All the obtained data have been recorded in schedule of preventive maintenance after reassemble the parts of mechanical body. Lastly, velocity of each robot has been tested after preventive maintenance execution and all the data obtained have been recorded. All the methodology has been provided follow standard procedure created during developing a framework of robot soccer preventive maintenance.

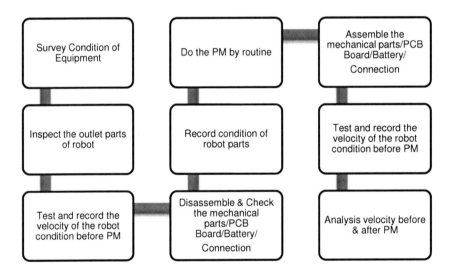

Fig. 1. Flow chat methodology of research

4   Propose Preventive Maintenance Framework

Preventive maintenance actually contributes a long term benefits to Robot Soccer system especially to improved system reliability. The original velocity will be contributed of performance of robot during its operational. The essential benefit including decrease cost replacement of parts and decrease of system downtime. Developing a framework for preventive of robot soccer classified in three categorize: Replacements parts, Expertise and Clean and Oiling. The body of robot as model-specific will be categorizing respectively. There is issue or problem for each categorize produced by robot soccer mirosot. The main problem for the replacements parts is lack of movement for robot during it's operational. The velocity of the robot is under the optimum level to chase or during dribbling the ball because damage or bad condition parts.

Environmental of robot mirosot also influence performance of velocity of robot. The dusty or rusty parts of body robot will decrease performance of robot. All the parts need to clean and clear to give the robot in optimum performance and increasing their movement and velocity during dribbling and chase the ball. The expertise personnel come with technical knowledge also the main issue that must be consider in developing the frameworks of preventive maintenance. Executing standard procedure of preventive maintenance in good manner and follow the right step will produce a great performance for the robot soccer mirosot.

Fig.2 is the proposed of preventive maintenance procedure for each category based on the issue or problem occurred during developing the framework.

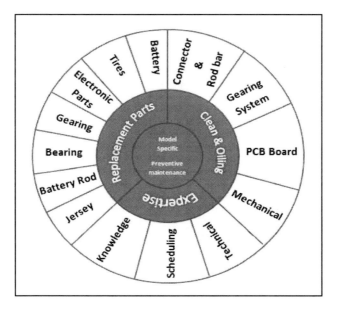

Fig. 2. Framework model-specific of Preventive maintenance

5 Propose Standards Procedure of Preventive Maintenance

Standard procedure of preventive maintenance of Robot Soccer model categorized in three parts: Mechanical body, electronic board and power source (battery). In Fig.3, a general standard preventive maintenance execute for the robot soccer mirosot for quarterly schedule.

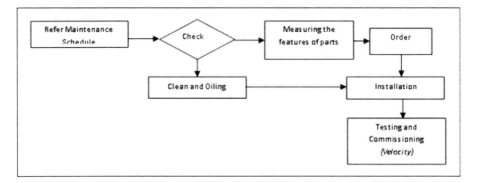

Fig. 3. General of Procedure Preventive maintenance

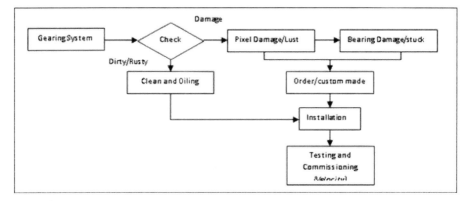

Fig. 4. Preventive Maintenance Procedure for Mechanical Body (Gearing System)

Mechanical parts are a main compartment in robot soccer mirosot that located a PCB board, gearing system, rod, battery, jersey for vision system. A balanced body of robot will be producing a great performance to the velocity of robot soccer mirosot. So, a proper preventive maintenance to the mechanical parts is a vital preservation to the robot soccer. The main parts of mechanical body concentrate on gearing system because of movement the robot rely on these parts. Fig.4 is standard in executing preventive maintenance for gearing system.

Fig.5 had shown standard procedure for preventive maintenance of PCB board of electronic circuit. The value of voltage must be enough to make sure the robot in optimum operational. All the corrosion or rusty port terminal must be clean and oiling from more bad result.

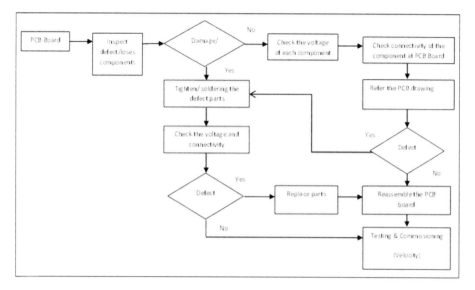

Fig. 5. Preventive Maintenance Procedure for PowerSource (Battery)

6 Results

Experimental analysis for the velocity of robot execute in developing framework of preventive maintenance. The distance at field of robot soccer mirosot measured and duration from first point recorded until the second point. Fig.6 had shown the distance for the experimental analysis.

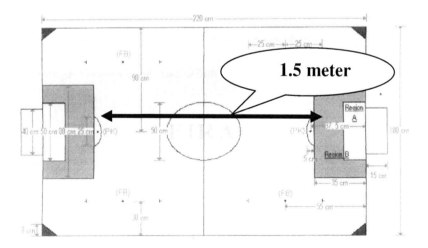

Fig. 6. Distance for measuring the distance for velocity experiment

The experiment data for quarterly scheduling preventive maintenance use for calibrate the velocity of the robot soccer mirosot. Fig.7, Fig.8 and Fig.9 had shown comparison before preventive maintenance result before and after executing a preventive maintenance.

Table 2. Result for velocity of the robot 1 (Goalie) before and after preventive maintenance

PM Schedule (every 4 month)	1		2		3		4	
	Before	**After**	Before	**After**	Before	**After**	Before	**After**
Velocity	0.679	**0.711**	0.612	**0.647**	0.503	**0.568**	0.688	**0.698**
Condition of Bearing	Dusty	o.k	Dusty	o.k	bad	bad	changed	changed
Condition of Tires	Dusty	o.k	bad	bad	bad	bad	changed	changed

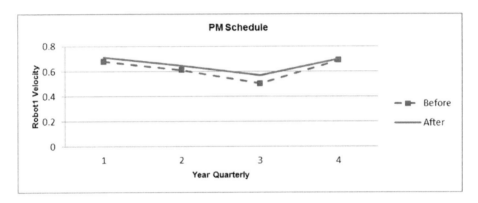

Fig. 7. Comparison between before and after preventive maintenance for robot 1 (Goalie)

Table 3. Result for velocity of the robot 2 (Sweeper) before and after preventive maintenance

PM Schedule (every 4 month)	1		2		3		4	
	Before	**After**	Before	**After**	Before	**After**	Before	**After**
Velocity	0.670	**0.688**	0.620	**0.638**	0.510	**0.545**	0.698	**0.698**
Condition of Bearing	Dusty	o.k	Dusty	o.k	bad	bad	changed	changed
Condition of Tires	o.k	Dusty	bad	bad	bad	bad	changed	changed

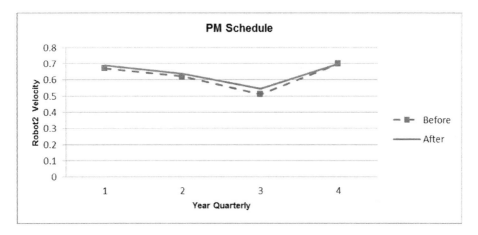

Fig. 8. Comparison between before and after preventive maintenance for robot 2(Sweeper)

Table 4. Result for velocity of the robot 3 (Attacker) before and after preventive maintenance

PM Schedule (every 4 month)	1		2		3		4	
	Before	**After**	Before	**After**	Before	**After**	Before	**After**
Velocity	0.649	**0.676**	0.524	**0.588**	0.484	**0.54**	0.647	**0.664**
Condition of Bearing	Dusty	o.k	Dusty	o.k	bad	bad	changed	changed
Condition of Tires	Dusty	o.k	bad	bad	bad	bad	changed	changed

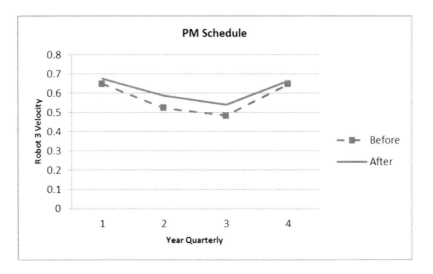

Fig. 9. Comparison between before and after preventive maintenance for robot 3 (Attacker)

Hypothesis Pair 1:

H_o: Velocity increase after conducting preventive maintenance for robot 1
H_a: Velocity not increase after conducting preventive maintenance for robot 1

Hypothesis Pair 2:

H_o: Velocity increase after conducting preventive maintenance for robot 2
H_a: Velocity not increase after conducting preventive maintenance for robot 2

Hypothesis Pair 3:

H_o: Velocity increase after conducting preventive maintenance for robot 3
H_a: Velocity not increase after conducting preventive maintenance for robot 3

Ho = Null Hypothesis, Ha = Alternative Hypothesis , μ = mean

We conducted a t-test for paired samples to analyse the significance of everyrobot velocity before and after preventive maintenance have done.

The t-test for paired samples (also referred to as a t-test for correlated groups) is used to compare the means of two samples of scores when there is a logical basis for connecting each score in one sample with a particular score in the other samples [10]. From the result, the significance value for both pair 1 and pair 2 are less than 0.05, thus the null hypothesis are rejected. The pair 1 of the t-test indicates that the difference between the mean of velocity value of the robot before and after preventive maintenance is significant and not coincident with $p < 0.05$. The result of the pair 2 and pair 3 also shows that robot velocity value of the robot before and after preventive maintenance is significant and not coincident with $p < 0.05$.

7   Conclusion

The effectiveness of a preventive maintenance should be evaluated in long term evaluation. The beneficial of preservation from the parts damage must be one primary objective in executing the preventive maintenance. The framework for model-specific of preventive maintenance should be experimented in long duration to obtain the realistic result of effectiveness of the preventive maintenance. Combination of preventive maintenance features like replacements part, expertise and oiling and clean also generate a good result for effectiveness if preventive maintenance.

The right solution for the preventive maintenance should be evaluated in different perspective. In this paper, we try to choose a model-specific as our goal in developing a framework of preventive maintenance. Finally, the framework can be applied to robot soccer model in preservation from the damage and decrease of costing.

8   Future Works

The framework of preventive maintenance presented in this paper was contributed to model-specific of robot soccer application. Other implementation in future research might be continuing with another advance application to improve the system. There are also different approach and characteristic to upgrade and enhance the reliability of robot soccer usage. Automationprocedure that used now day in industrial might be implemented to robot soccer system in executing preventive maintenance. The system might be automated if all the characteristic of artificial intelligence implement in preventive maintenance works.

This paper also not consider the another parts of robot soccer system such as transmission system, vision system, AI system and etc to execute preventive maintenance. In future, all the features might be considered to make robot soccer system is a beneficial research to explore a new technology.

References

1. Kim, J.-H., Kim, D.-H., Kim, Y.-J., Seow, K.-T.: Soccer Robotics. Springer, Heiderberg (2004)
2. Chen, C.-T.: Dynamic preventive maintenance strategy for an aging and deteriorating production system. Ta-Hwa Institute of Technology, Hsinchu 307, Taiwan (2010)
3. Al-Mishari, S.T.: Modelling Preventive Maintenance for AuxiliaryComponenets. Emerald Group Publishing Limited (2008)
4. Roux, O., Duvivier, D., Quesnal, G., Ramat, E.: Optimizing of Preventive Maintenance through combined maintenance-production simulation model (2010)
5. Nosoohi, I., Hejazi, S.R.: A multi-objective approach to simultaneous determination of spare part numbers and preventive replacement times. Applied Mathematical Modelling (2011)
6. Bartholomew-Biggs, M., et al.: Modelling and optimizing sequential imperfect preventive maintenance. Reliability Engineering and System Safety, 53–62 (2008)
7. Selvik, J.T., Aven, T.: A framework for reliability and risk centered maintenance. Reliability Engineering and System Safety, 324–333 (2011)
8. Mann, L., Saxena, A., Knapp, G.M.: Statistical-based or condition-based preventive maintenance? Journal of Quality in Maintenance 1(1), 45–59 (1995)
9. Alardhi, M., Hannam, R.G.: Preventive Maintenance Scheduling for multi-cogeneration plants with production constraints. Journal of Quality in Maintenance Engineering 13(3), 276–292 (2007)
10. Stern, L.D.: A Visual Approach to SPSS for Windows. A Guide to SPSS 15.0. Pearson Education Inc., London (2008)

Optimal Features and Classes for Estimating Mobile Robot Orientation Based on Support Vector Machine

Zainal Fitri Mohd Zolkifli[1], Mohamad Farif Jemili[1],
Fadzilah Hashim[1], and Siti Norul Huda Sheikh Abdullah[2]

[1] Department of Information Technology and Communication
Politeknik Sultan Abdul Halim Mu'adzam Shah
06000 Bandar Darulaman, Jitra, Kedah, Malaysia
zainal.fitri@gmail.com, farifmal@yahoo.co.uk,
fadzilah69@yahoo.com.my
[2] Center for Artificial Intelligence Technology
Faculty of Information Science and Technology,
Universiti Kebangsaan Malaysia,43600, Bangi,Selangor, Malaysia
mimi@ftsm.ukm.my

Abstract. In order for a mobile robot to perform its assigned tasks, it often requires a representation of its environment such as knowledge of how to navigate in its environment, and a method for determining its position in the environment. A major problem in computer vision and machine learning is to achieve a good feature as it can largely determine the performance of a vision system. A good feature should be informative, invariant to noise or a given set of transformations, and fast to compute. Also, in certain settings sparsity of the feature response, either across images or within a single image, is desired. Our objective of this paper is to obtain optimal features as well as determining the optimal class of angle in order to estimate mobile robot orientation single or unified images from two camera orientations. We introduce feature selection process before classifying features based on support vector machine classifier. We achieve better accuracy rate by only reducing its feature number from 30 features down to only 17 features on unified images. Furthermore, we also find that only 5 classes of robot angles are sufficient to estimate robot orientation correctly.

Keywords: image calibration, robot soccer, Support Vector Machine, feature reduction, mobile robot orientation.

1 Introduction

Robot soccer is a soccer game which its program consists of calibration skills, communication and strategy modules. Even though, this new developments have been conducted and designed on this game program, there are still remain many issues still remained unsolved such as getting the positions of camera, optimizing colour detection in robot players and etc. Before starting the game, each team needs to find the most optimum parameter tunings in image calibration stage. This phase is the hardest part because it will reflect the whole performance of the robot players.

T.-H.S. Li et al. (Eds.): FIRA 2011, CCIS 212, pp. 270–279, 2011.
© Springer-Verlag Berlin Heidelberg 2011

In image calibration at the first stage of the phase, it is tedious step but it requires special attention to avoid from failure in robot strategies. Many research attempted various manual or automatic methods to find robot orientation tuning such as Paulraj et. al [1,2] using neural network. Neural Networks (NNs) is one of the branches of Artificial Intelligence (AI). It is also known as connectionist systems, parallel distributed processing (PDP), neural computing, and artificial neural systems. NNs try to mimic how a brain and nervous system works. NNs are famous for their inductive learning ability, i.e., the ability of learning through experience. The performance of Neural-Network (NN)-based classifiers is strongly dependent on the data set used for learning. In practice, a data set may contain noisy or redundant data items. Thus, feature selection is an important step in building an effective and efficient NN-based classifier. For example, Ackerman and Itti [3] system used a trained feed forward neural network to estimate four features from global or coarsely localized image spectral information. The results showed that much intuitively navigationally useful information could be extracted from such computationally inexpensive analysis [3].

The objective of this paper is to determine the optimal features reduction for estimating the mobile robot orientation and to determine the optimal class of angle for mobile robot. Our proposed method used data set extracted from the previous experiment using NNs [1].

We organize our paper into five sections. The first and second sections explain on introduction and basic state of the art in robot soccer domain subsequently. Then, we discuss on our proposed method for optimal features reduction and classes of angle for estimating the mobile robot orientation in section three or methodology. Pertaining to above, we evaluate our proposed framework with the state of the art method in section experimental results and discussion or fourth section. Finally, we conclude our findings in section five.

2   State of the Art

According to Paulraj et. al [1], robot orientation prediction can also be solved by applying local and global features of single or multiple overhead cameras and neural network as the classifier. They adjusted the camera at two orientations such as 90° and 22.5°. At the same time, they also suggested unified image by superimposing two images from two cameras and applying 'or' operator. They achieved from 81% to 91% of accuracy rate when predicting the camera orientation and found that by using unified images, the overall performance can be increased. On the other hand, this proposed method requires additional cost of hardware and time consuming.

Another study was conducted by Donald Bailey [2] in order to find the best way to determine the accuracy of the mobile robot using typical vision system. The studies include relative estimation of the camera position to the robot soccer field by purely imaging the field and back projecting the parallax offset of the walls. The geometrical features were used are from 4 to 5 features which included local and global features from different camera angle. The study has been introduced as the earlier methods to determine camera position based on calculation of the robot soccer field.

In other situation, we study about image analysis using Zernike and Legendre polynomials. This technique is defined only inside the unit circle and coordinate transformation is required to compute the moments[4]. This coordinate representation does not easily yield translation invariant functions, which are also sought after in pattern recognition application [5]. According to Thawar Arif et al [4], three kinds of moments (Geometric, Zernike and Legendre) and Nearest Neighbor Classifier are used. Various features had been used General features and Domain-specific features.

In Geometric moments the shape of an object is an important because it represents the essential characteristic of an object. The Zernike moments is used to overcome the shortcomings of information redundancy present in the popular geometric moments [6].Moments with Legendre polynomials as kernel function, denoted as Legendre moment [6].   This moment is achieved by using a combination of corresponding invariant of geometric moments. They also used Neighbor classifier to relies on a metric or distance function between 4 points. This classifier is used to compare the feature vectors and stored in the database ETH-80.   The sample consists of 3280 images and resized into 60X60 pixels. The result is to recognize the images of 3D objects using Nearest Neighbor classifier based on the Legendre moments is higher. Based on previous work, there are still rooms for improving manual or automatic camera orientation prediction or calculator.

3   Methodology

Prior to above discussions, we propose a revised framework for estimating the mobile robot orientation via camera orientation, feature selection, angle classes determination and support vector machine as in Fig. 1. It consists of three stages such as image capturing, global and local feature analysis and Support Vector Machine classifier. We detail out the following subsections.

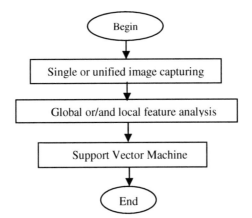

Fig. 1. The proposed framework of the mobile robot orientation prediction based on global and local feature analysis and support vector machine

A. *Image Capturing*

The images are captured at two different orientations such as 90° and 22.5° [7]. Based on previous experiments [1], these orientations are selected to obtain better human view. Fig. 2 is an example illustration of the camera orientations.

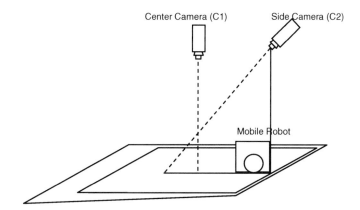

Fig. 2. Camera orientations at different angles (a) 90° and (b) 22.5° [1]

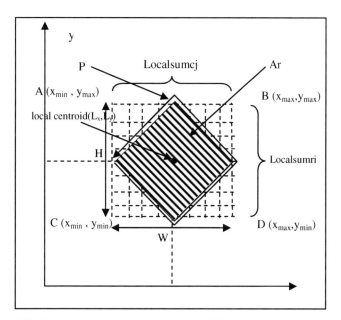

Fig. 3. Images captured from single or unified image camera [1]

The second camera (22.5 °orientation) point of view is skewed towards the end of the field where the camera is facing thus changing the centre of the field. A total of 107 images are captured from each camera at different orientation varies from 0° to 90°. Four coordinates are captured which are A = (x_{min} , y_{max}), B = (x_{max}, y_{max}), C = (x_{min} , y_{min}), and D = (x_{max}, y_{min}) of the robots to localize the mobile robot as Fig. 3:

B. Global and local feature analysis.

We use the similar features as proposed by Paulraj et. al [1]. The features obtained are 10 features from camera1, 10 features from camera 2 and 30 features from unified image. For camera 1 and 2, the features parameters are as shown below in Table 1:

Table 1. Feature Extraction Parameters

No	Feature name	Feature notation	Number of feature before FS
$X_{1,2}$	global centroid	(G_x, G_y)	2
$X_{3,4}$	local centroid	(L_x, L_y)	2
X_5	height	H	1
X_6	width	W	1
X_7	sum of local columns	Localsumcj	1
X_8	sum of local rows	Localsumri	1
X_9	area	A	1
X_{10}	perimeter	P	1
	Total features		10

While for unified images, 10 features derived from camera 1, 10 features from camera 2 and 10 features derived from superimposed image of camera 1 (C1) and 2 (C2) for a total of 30 features.

We use the orientation derived from the output of NNs and divide the output into 9 classes, 6 classes and 5 classes in order to train the features for classification. This is done by increasing the range of data from 10, 15 and 18 respectively. We also investigate the significant of with and without feature selection process towards Paulraj et. al 's feature introduction [1].

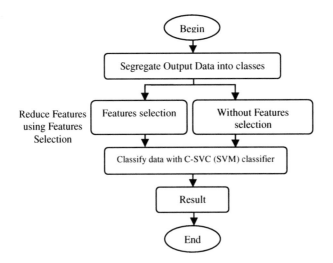

Fig. 4. Features analysis/selection step by step

We apply Classifier Subset Evaluator which uses a classifier to estimate the 'merit' of a set of features introduced by Paulraj et. al [1]. In this case, we use sequential minimal optimization (SMO) [8] to estimate the accuracy of the features whereby the search method used is a simple genetic algorithm as described in [9].

The following figure shows the steps involved in features selection (FS):-

C. Classification

We perform classification using Support Vector Machine precisely one versus one method namely C-SVC. The Support Vector Machine is formulated as below:

Let us assume that $\{X_1, ..., X_n\}$ are our data set of feature parameters as Table 1 for respective C1, C2 and Unified condition, and let $Y_i \in \{1, -1\}$ be the class label of X_i where either 5, 6 or 10 classes of angle in the range from 0 until 90°. We concern on decision boundary should classify all points correctly as below:

$$Y_i(W^T X_i + b) \geq 1, \forall i \tag{1}$$

We optimize W^T based method SMO method and use Gaussian kernel function to indicate the class sector [10]. The SVM classifier is trained with a set of samples to find the optimized parameter such as C and γ. Prior to that, we normalize our data using logarithmic scaling. The classification performed with 67% training data set and the remainder is used for test. The optimized parameter is determined through a series of testing (more than 10 times) regarding to the dataset with and without feature selection.

4   Experimental Results and Discussion

The results shows increased in accuracy of correctly classified instances using optimal classes of angle and features suggested by Platt et. al. [8,11].

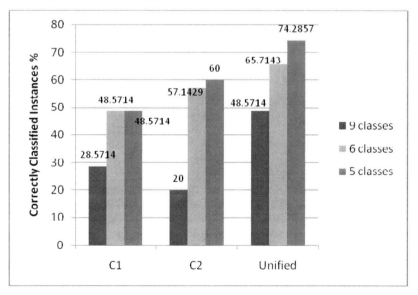

Fig. 5. Correctly classified instances in percentage without Feature Selection and Optimal Classes based Support Vector Machine

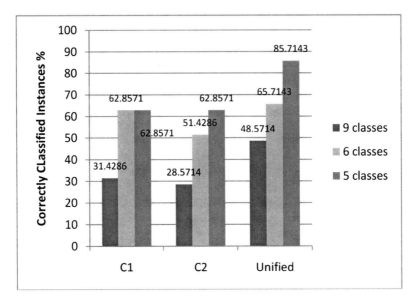

Fig. 6. Correctly classified instances in percentage with Feature Selection and Optimal Classes based Support Vector Machine

Table 2. Chosen features after performing features selection for C1

C1		
9 classes	**6 classes**	**5 classes**
globalcentroidx	globalcentroidy	globalcentroidx
localcentroidx	localcentroidx	globalcentroidy
LocalCj	localcentroidy	localcentroidx
Localri	LocalCj	localcentroidy
localsumcj	Localri	LocalCj
localsumri	localsumri	Localri
area	area	localsumri
perimeter		perimeter

From the charts above, there are three types of output from camera 1 (C1), camera 2 (C2) and unified image (Unified) and these output are tested using feature selection and without feature selection. The bar chart represents each output from the percentage that we extracted from the correctly classified instances.

The accuracy increment can be seen as we decrease the number of classes from 9 classes to 6 and lastly 5 classes. It shows that by using optimal classes of data does increase the percentage of accuracy in term of classification. We also compare the trend using results that we performed with features selection and followed by without features selection. The trend shows increasing of percentage for C1, C2 and Unified.

Feature selection step is important in determining the essential features for the gathered features for a dataset. After applying features selection for C1, C2 and Unified, chosen features are obtained as per Table 2, 3 and 4.

Table 3. Chosen features after performing feature selection for C2

C2		
9 classes	**6 classes**	**5 classes**
globalcentroidx	globalcentroidx	globalcentroidx
globalcentroidy	globalcentroidy	localcentroidx
localcentroidy	localcentroidy	localcentroidy
LocalCj	LocalCj	LocalCj
Localri	Localri	Localri
localsumcj	localsumcj	localsumcj
localsumri	localsumri	area
	area	perimeter

Table 4. Chosen features after performing feature selection for Unified

Unified		
9 classes	**6 classes**	**5 classes**
C1globalcentroidy	C1globalcentroidx	C1LocalCj
C1localcentroidx	C1globalcentroidy	C1Localri
C1localcentroidy	C1localcentroidy	C1localsumcj
C1LocalCj	C1LocalCj	C1localsumri
C1localsumcj	C1Localri	C1area
C1localsumri	C1localsumcj	C1perimeter
C1area	C1localsumri	C2globalcentroidy
C2globalcentroidx	C1area	C2LocalCj
C2localcentroidx	C1perimeter	C2Localri
C2localcentroidy	C2globalcentroidx	C2localsumcj
C2LocalCj	C2globalcentroidy	C2localsumri
C2Localri	C2localcentroidx	C2perimeter
C2localsumcj	C2localcentroidy	Uglobalcentroidx
C2localsumri	C2LocalCj	Ulocalcentroidx
C2perimeter	C2Localri	ULocalri
Uglobalcentroidx	C2localsumcj	Uarea
ULocalCj	C2localsumri	Uperimeter
ULocalri	C2area	
Ulocalsumcj	C2perimeter	
Ulocalsumri	Ulocalcentroidy	
Uarea	Uarea	
	Uperimeter	

Another comparison that we make is in term of performance of the classification where we measured time taken to build the model. Average time taken to build the model is measured with/without features selection and compared by classes.

Table 5. Average time taken to build the model with C1

	9 classes	6 classes	5 classes
Without Features Selection	0.0822	0.0538	0.0396
With Features Selection	0.0724	0.0452	0.0392

Table 6. Average time taken to build the model with C2

	9 classes	6 classes	5 classes
Without Features Selection	0.0952	0.0528	0.0386
With Features Selection	0.0784	0.0502	0.043

Table 7. Average time taken to build the model with Unified Images

	9 classes	6 classes	5 classes
Without Features Selection	0.1348	0.0912	0.0816
With Features Selection	0.1176	0.0732	0.0538

The value shown on the table 5, 6 and 7 are based on average of 50 experiments. The values shown by reducing the classes give a detrimental trend on all three types of images. The average time taken to build the model tested with features selection also shows reduction on all images.

5 Conclusions

Based on our research, we proposed optimal features for estimating mobile robot orientation using a feature selection method namely Classifier Subset Evaluator through SMO classifier and Genetic search method to evaluate the essential features and discarded the rest. This is done by classifying the data set into 9 classes, 6 classes and 5 classes. Based on our findings, the optimal features reduction shows increased both in term of accuracy and performance. By determining the optimal classes of data, we also achieved higher accuracy rate and increased performance in line with the optimal features reduction.

As conclusion we can conclude our findings that optimal features reduction and classes of angle for mobile robot orientation plays a part for the successful vision system in robot soccer in term of performance time and accurate estimation of mobile robot orientation. For future work, it is proposed to use different classifier and attribute selection algorithm.

Acknowledgement. The authors would like to give a lot of appreciation for the financial and motivation support from Polytechnique Department, Malaysia Higher Education Ministry and last but not least Image Recognition Team from Faculty of Technology and Information Science, UKM.

References

[1] Paulraj, M.P., Ahmad, B., Hema, C.R., Hashim, F.: Estimation of Mobile Robot Orientation Using Neural

[2] Bailey, D., Gupta, G.S.: Automatic Estimation of Camera Position in Robot Soccer. In: International Machine Vision and Image Processing Conference, IMVIP 2008, September 3-5, pp. 91–96 (2008)

[3] Ackerman, C., Itti, L.: Robot Steering With Spectral Image Information. IEEE Transactions on Robotics 21 (April 2005)

[4] Arif, T., Shaaban, Z., Krekor, L., Baba, S.: Object Classfication Via Geometrical, Zernike and Legendre Moments

[5] Chonga, C.-W., Raveendrab, P., Mukudan, R.: Translation invariants of Zernike moments

[6] Teague, M.: Image analysis via the general theory moments. J. Opt Soc. Am. 70(8) (1980)

[7] Khusairi Osman, M., Yusoff Mashor, M., Rizal Arshad, M.: An Optimized Camera-Object setup for 3D Object Recognition system

[8] Platt, J.: Fast Training of Support Vector Machines using Sequential Minimal Optimization. In: Schoelkopf, B., Burgeand, C., Smola, A. (eds.) Advances in Kernel Methods - Support Vector Learning (1998)

[9] Xu, D., Youfu, L., Min, T.: A Visual Positioning Method Based on Relative Orientation Detection for Mobile Robots. In: 2006 IEEE/RSJ International Conference on Intelligent Robots and Systems, October 9-15, pp. 1243–1248 (2006)

[10] Sheikh Abdullah, S.N.H., Omar, K., Marzuki, K.: License Plate Recognition using Support Vector Machine. In: International Conference on Electrical Engineering and Informatics 2009, Bangi, Malaysia, August 5-8, pp. 78–82 (2009)

[11] Govindaraju, V., Shi, Z., Schneider, J.: Feature extraction using a chaincoded contour representation of fingerprint images. In: Kittler, J., Nixon, M.S. (eds.) AVBPA 2003. LNCS, vol. 2688. Springer, Heidelberg (2003)

Multiple Robots Coordination and Shooting Strategy in Robotic Soccer Game

Awang Hendrianto Pratomo, Anton Satria Prabuwono,
Siti Norul Huda Sheikh Abdullah, and Mohamad Shanudin Zakaria

Center for Artificial Intelligence Technology (CAIT),
Faculty of Information Science and Technology, Universiti Kebangsaan Malaysia,
43600 UKM Bangi, Selangor D.E., Malaysia
awang.upn@gmail.com, {antonsatria,mimi,msz}@ftsm.ukm.my

Abstract. In this paper, we create a passing, obstacle avoiding and shooting strategy for robotic soccer coordination. Based on a predefined-scenario in a robotic soccer game, we simulate a mini case study which involves two robots and a ball. We modify role, act and behavior method to meet the game requirements. About 61% of the testing achieved the shooting of a goal by manipulating and redesigning the strategy. The shortest goal shooting time was about 5 seconds. We hope to improve this initial strategy in the future.

Keywords: Multiple robots coordination, obstacle avoidance, shooting strategy.

1 Introduction

Robotic soccer is a game similar to a football match. The only one significant difference in a robotic soccer game is that the players are not humans but are a group of robots. Normally, these robots are miniature wheeled mobile vehicles which can be mechanically constructed and easily controlled [1]. Each team consists of several players, one of them is the goal keeper. These robots are autonomous and can be controlled by computer. A CCD camera is mounted overhead of the field and it is connected to the computer whereby all the information is captured in real-time i.e. the position of individual (team and opposition) players, the ball and the grid of the soccer field. Each team has its own camera and computer and the robots are playing independently or without human intervention. A human referee will supervise and monitor throughout the game. From another perspective, this mini game can trigger or initiate many learning issues. The robotic soccer game involves strategies and cooperation between players in order to shoot as many goals as possible into the opposition goal and to defend any goal scoring from the opposition players within a given time [2]. Normally, the winning team is the team that has scored more goals.

The robotic soccer team must undergo an intensive learning process about the robotic soccer game. At first, they must acquire some low-level skills that allow them to manipulate the ball. Examples of the low-level or the basic skills of the robotic soccer game are dribbling and controlling the ball. Their strategies must be robust and well-suited according to the opponents' behaviour. Each robot must be adaptable to

T.-H.S. Li et al. (Eds.): FIRA 2011, CCIS 212, pp. 280–289, 2011.

the new situations and opponents' behaviour [3]. Besides, each team player must also be able to cooperate in passing and shooting the goal.

Several studies have been conducted in developing robotic soccer games [2], [4]. However in practice some of the robot strategies are impractical because they are less dynamic, lack movement and there is a shortage of cooperation strategies based on various environments and conditions. Therefore, they shifted to develop new strategies using an intelligent agent approach [5].

The contribution of this paper is to develop and compare the best two robot coordination strategies in robotic soccer. We describe the case as below:

Step 1: Pass the ball from the first robot to the second robot.
Step 2: The second robot avoids the obstacle and shoots the ball into the goal.

This paper is organized as follows. Section 2 explains state of the art. Section 3 describes the proposed strategies. Experimental results and discussions are shown in section 4, and section 5 presents our conclusions.

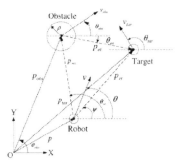

Fig. 1. The velocity control of the robots' movements

2   State of the Art

In the robotic soccer team, five autonomous robots are used. In this game, humans can control and coordinate these autonomous robots. Only one robot is controlled using a joystick while the remainders are controlled by a series of strategies [5]. This method of controlling will create a human control system which allows the autonomous robots to collaborate among themselves. A research on the fuzzy rule decision making mechanism has been done on autonomous robots [6]. They claim that this rule-based approach is very similar to the MiroSot robots. Unlike the MiroSot robots, their approach enables the robot to make decisions and to act autonomously. In other words, the decisions are based on local sensor data of the robot instead of global data.

Another research which concentrates on mobile robot path planning is described using the potential field method to speed up the mobile robot planning. This approach was still able to track a moving object even though there were moving obstacles in surroundings [7]. They kept the potential function in its traditional way, and derived an explicit function for controlling the robot speed at a certain stage such as the speed of the target (obstacle) and relative positions between the obstacles, the goal and between the robots themselves. The speed of the robot function can be determined as following (refer to Fig. 1).

$$\| v \| = \left[\left(\| v_{tar} \| \cos(\theta_{tar} - \psi) - \sum_{i=1}^{n} \beta_i \| v_{obs_i} \| \cos[(\theta_{obs_i} - \theta_{ro_i}) + \xi_1 \| p_{rt} \|) \right]^2 + \right.$$
$$\left. \| v_{tar} \|^2 \sin^2(\theta_{tar} - \bar\psi) \right]^{\frac{1}{2}} \tag{1}$$

Where:

v = velocity of the robots

v_{obs_i} = velocity of the object

v_{tar} = velocity for the robots to track the object

θ_{tar} = angle of v_{tar}

θ_{ro_i} = angle of p_{ro}

ψ = angle of the relative position from the robot to the target

ξ_1 = scaling factor

p_{rt} = the relative motion between the robot and the target

$\| v_{obs_i} \|$ and θ_{obs_i} are the speed and direction of i_{th} obstacle

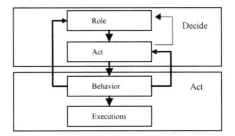

Fig. 2. Control architecture in robotic soccer

Reference [8] examines the realization and visualization of ball passing, and move-to-ball behavior of a robotic soccer game. This research describes three identical mechanical mobile robots with a formation ready to pass a ball cyclically in a zigzag pattern. In their work, a control command driven mobile robot motion simulator with a controller and dynamics of mobile robots is built. In this simulator, kick motion follows a physical law, and a simplified collision check and response model is utilized for the efficient detection of the hitting of a robot with the ball or other robots. The realization of specific ball passing strategy to drive each soccer robot in a position to receive a pass included three levels of organization, coordination, and execution, integrated design of a dynamic formation and role change scheme, ball position estimation, and coordinated trajectory planning and tracking control.

The basic control architecture of robotic soccer is shown in Fig. 2 [9]. This architecture describes the process of decision-making and action in the game. Each process produces behavior based on its own strategy although it may rely on the presence and operation of players [10].

In addition, defining the behavior of the actual football match into robot soccer strategy is not an easy task. It will also include modeling and setting strategy of the opponent's behavior. MiroSot system usually uses hard coded sets formally known as if-then rule [2], [4], [5]. This approach is adequate to perform simple reasoning. However it is a bit troublesome when updating, adding or deleting the rules.

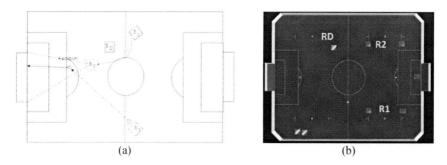

(a)   (b)

Fig. 3. (a) Position and moving path of R_1, R_2, RD and R'_2, (b) simulation view before the competition starts. The actual positions of the robots R_1, R_2 and R_3 and the ball.

From the previous research, it can be concluded that, robotic soccer presents an autonomous control structure. The planning and activities of each robot in the team is mostly controlled by the computer systems [9]. The robots must able to think fast and make decisions in a few milliseconds. In this research, the main objective is to test the results of passing and shooting strategies in robotic soccer, a case which is shown in Fig. 3(a) and 3(b).

3 The Proposed Strategy

Normally, the robotic soccer game is equipped with a camera mounting overhead and a personal computer for each team at the side of the field. The camera captures the information of all objects on the field such as robots, the ball and the field lines. The personal computer which embeds the image processing module, will also act as the controller. It reads the image data, analyzes them and detects the position of each object on the field. We apply Visual C++ programming to create strategies for each player.

The mission is to coordinate a passing and shooting strategy. There are three robots namely R_1, R_2 and RD. R_1 and R_2 are the team players while RD is the opponent robot. R_1 passes the ball to R_2 but R_2 must avoid the obstacle, called RD in advance. Then, R_2 shoots the ball into the goal. Fig. 3(b) shows the simulation view of the actual positions of the R_1, R_2, RD robots, and the ball. The R_1 and R_2 robots are placed in parallel. The opponent robot, RD is located in front of R_2. RD acts as an obstacle in this game.

We employ a similar strategy to [11] and organize them into four levels. Firstly, the role level of the control architecture is related to the **who** issue. We choose a role and area of maneuver in the playground for each robot such as defender, midfielder, attacker, and goalkeeper. Secondly, in the act level that associates with the **what** issue, we select appropriate actions for each robot pertaining to the objective role. Some examples of actions are dribbling, passing, shooting, and blocking. Thirdly, in the behavior level, we simulate the **how** issue. This refers to the prior condition checking for each robot. For example, the robot moves left or right upon knowing the current scenario such as the obstacle existence. Fourthly, in the execution level, we correlate with the **when** issue. Prior to the above, we estimate the precise time and motion for each robot. This level involves the hardware control such as the robot motor for each movement type.

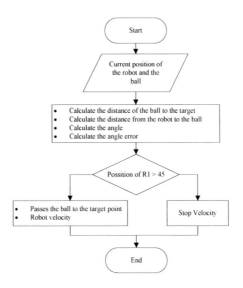

Fig. 4. The proposed passing strategy flowchart

Pertaining to above structure, we construct two algorithms for passing and shooting. Both of these algorithms execute simultaneously. We describe the proposed passing algorithm (Fig. 4) as below.

Algorithm 1. The proposed passing algorithm
Input: Current positions of all robots and the ball.
Output: Robot velocity.
Step 1: We assume each robot position as (x_{Rn}, y_{Rn}) where R is the robot and $n = \{0 \ldots 2\}$, the ball as b, and the target position as (x_t, y_t). We calculate the distances of the ball to the target, based on each x and y coordinate as the following.

$$(dx_{tb}, \ dy_{tb}) = (x_t - x_b, y_t - y_b) \tag{2}$$

Step 2: We calculate distance, from the robot to the ball as the following.

$$(dx_{Rn\,b}, \ dy_{Rn\,b}) = (x_{Rn} - x_b, y_{Rn} - y_b) \tag{3}$$

Step 3: We estimate the rotation angle of the ball to the target, α and the rotation angle of the robot to the ball, β as below.

$$\alpha = {180}/{\pi} * atan^2(dx_{tb}, \ dy_{tb}) \tag{4}$$

$$\beta = {180}/{\pi} * atan^2(dx_{Rn\,b}, \ dy_{Rn\,b}) \tag{5}$$

We calculate the desired shooting angle, θ_{bt} of the ball to the target point.

$$\theta_{bt} = 2 * (\alpha - \beta) \tag{6}$$

Step 4: We calculate the angle error as the following.

$$\theta_e = \theta_{bt} - \theta_{Rn} \tag{7}$$

where θ_{rn} = the robot rotation.

Step 5: We set the kinematic value or propotional gain, K based on θ_e .
Step 6: We calculate the velocity of the robot to move left, vl_{Rn} or right, vr_{Rn} as below.

$$vl_{Rn} = vc - K \times \theta_e \tag{8}$$

$$vr_{Rn} = vc - K \times \theta_e \tag{9}$$

where vc = velocity control of the robot heading direction.
Step 7: End.

In addition, the proposed passing flow chart is depicted as shown in Fig. 4. After R_1 has passed the ball to R_2 using the above passing strategy, we construct another strategy for moving with obstacle avoidance and shooting the ball. Based on the current positions of the robot, obstacle and the target ball (Fig. 5), we predict the robot velocity and directions to the target position. Let us say, if the opponent robot (obstacle (x_o, y_o)) exists in between the current (x_{Rn}, y_{Rn}), the desired $(\breve{x}_{Rn}, \breve{y}_{Rn})$, and the target positions $(\bar{x}_{Rn}, \bar{y}_{Rn})$, the robot must avoid the obstacle if the perpendicular distance to the line, l of the obstacle, d is less than radius, r_o. The algorithm (Fig. 6) for shooting with or without obstacle avoidance is explained as algorithm 2 [12].

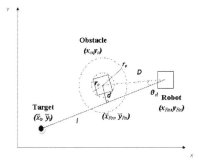

Fig. 5. The illustration of the position for each robot, obstacle and target ball

Algorithm 2. The proposed moving with obstacle avoidance and shooting algorithm
Input: Current position of all the robots and the ball, and the subsequent position of the robots.
Output: Robot velocity.
Step 1: We assume D as the distance of the robot to the obstacle, l is the length of the robot position (x_{Rn}, y_{Rn}), to the obstacle position, (x_o, y_o), $diff\theta$ is the angle difference between the robots to the obstacle and θ_d is the angle of the robots. We calculate the distance of the robots to the obstacle as the following equation:

$$D = \sqrt{(x_o - x_{Rn})^2 + (y_{Rn} - y_o)^2} \tag{10}$$

$$l = fabs(x_o - x_{Rn}) \times \sin(\theta_d) + (y_{Rn} - y_o) \times \cos(\theta_d)) \tag{11}$$

$$\theta_{Rn} = \text{atan}\,(y_o - y_{Rn}, x_o - x_{Rn}) \tag{12}$$

$$\text{diff}\theta = \theta_d - \theta_{Rn} \tag{13}$$

Step 2: We predict the subsequent action based on the condition of obstacle and target point, where δ is a safety margin for collision avoidance and r_o is the radius of the obstacle.

$$action = \begin{cases} \textit{move and avoid} & if \\ (D < r_o + \delta)\ and\ fabs\ (diff\theta < \frac{\pi}{2}) & \\ \textit{move and go to Step 6} & else \end{cases} \tag{14}$$

Step 3: If the action prediction is 'move and avoid', we calculate the distance from the center of gravity of the obstacle to the line, l as the following.

$$d = \frac{ax_o + by_o + c}{\sqrt{a^2 + b^2}} \tag{15}$$

Step 4: We calculate the radius of the robot and the relative position r_v to the obstacle, where r_{Rn} is the radius of the robot, using the following equation.

$$r_v = r_{Rn} + r_o \tag{16}$$

Step 5: We estimate the desired position of the robot, $(\check{x}_{Rn}, \check{y}_{Rn})$ and the bearing or the orientation of the robot B_{Rn} as below.

$$B_{Rn} = \begin{cases} \textit{Clock wise} & if\ (d \le 0) \\ \textit{Counter clock wise} & else\ (d > 0) \end{cases} \tag{17}$$

$$\check{x}_{Rn} = \frac{d}{|d|}\ y_{Rn} + x_{Rn}\ (r_v^2 - x_{Rn}^2 - y_{Rn}^2) \tag{18}$$

$$\check{y}_{Rn} = \frac{d}{|d|}\ x_{Rn} + y_{Rn}\ (r_v^2 - x_{Rn}^2 - y_{Rn}^2) \tag{19}$$

Step 6: We compute the direction of the robot behind the ball for moving towards a shooting position facing the goal.

$$\bar{\check{x}}_{Rn} = x_b - dx_{R_nb} \times \cos\,(\theta_{bt}) \tag{20}$$

$$\bar{\check{y}}_{Rn} = y_b - dy_{R_nb} \times \sin\,(\theta_{bt}) \tag{21}$$

Step 7: We use algorithm 1 to calculate the velocity of the robots.
Step 8: End.

The proposed shooting with obstacle avoidance flowchart is illustrated as shown in Fig. 6. Based on algorithm 1, R_1 passes the ball to the shooting area when it's x_{R1}coordinate reaches or equals to 45 as shown in Fig. 7(a). We assume that R1 has passed the ball to the target (x_t, y_t) position. Pertaining to algorithm 2, R2 simultaneously moves and stops until its y_{R2} coordinate reaches above 20 or the x_{R2} coordinate is in the position between 50 and 60. R_2 always keeps detecting the gap variance between itself and the ball while moving and avoiding the obstacle, RD. If the distance between the ball and robot R_2 is less than 45, R_2 will shoot the ball into the goal as shown in Fig. 7(b).

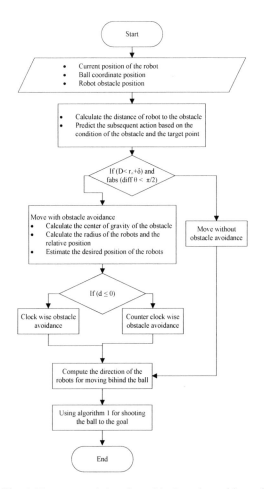

Fig. 6. The proposed shooting with obstacle avoidance flowchart

(a)   (b)

Fig. 7. (a) R_2 avoids RD by turning to left side while moving to R'_2, (b) R_2 shoots the ball into the goal successfully.

On the other hand, we also improve other multiple functions such as ball prediction, velocity, angle, positioning, kicking, waiting, and obstacle avoiding in order to achieve precise passing between robots and accurate shooting.

Table 1. The details for each task: T_1, T_2, T_3 and T_4 that have been marked in Fig. 3

Task	Process
T_1	R1 passes the ball to the R2
T_2	R2 moves towards the ball and avoids the obstacle
T_3	R2 shoots the ball to the goal
T_4	The ball when entering the goal or otherwise

4 Result and Discussion

From the case, four tasks had been outlined and each task was given between 1 and 5 scores. Each task is represented as T_1, T_2, T_3 and T_4 as shown in Table 1. We tested 5 strategies with 50 times of test runs for each strategy. In Table 2, we can see all strategies achieved to execute T_1 which is passing the ball from R_1 to R_2. However the task T_2 is decreased compared to T_1 because only several strategies could handle the obstacle avoidance as the proposed strategy. They could only manage to avoid the obstacle but they did not handle the speed and distance variances between the robot R_2 and the target position. We can also observe that T_2 is highly correlated to the achievement in T_3 and T_4. Furthermore, the probability of shooting the ball into the goal at an accurate position can also be reduced if the obstacle avoidance strategy is improper or incomplete. The best strategy is number 2 in which can make 37 goals. Furthermore strategy 5 produced 20 goals and 17 own goals.

On the other hand, estimating the precise execution time for each task is vital and critical. If each robot can predict all the possible tasks from the team or opponent players, and the ball, then they can compute the consequences or the subsequent actions to be handled. Based on this result, we also observed the execution time of each successful goal. Therefore, the avoidance strategy must also cover the cost and effect of each player's action.

Table 2. Result of strategy testing

Strategy No.	Goal	Own Goal
1	6	1
2	34	2
3	28	0
4	22	8
5	20	17

5 Conclusion

The robotic soccer strategy involves creating algorithms for passing and shooting the ball into the goal for that particular simulator. The aim is to improve the stability and velocity of the robot's behavior to kick the ball into the goal. Hence, using the experience gained in the game, the erroneous aspects which occurred in the previous algorithm was improved upon and understood as well. This strategy algorithm clearly distinguishes the different types of functions that have been executed and the effect of it. In order to achieve this aim, the respective positions need to be set up. This will enable the robot to ensure the success of passing and shooting.

In addition to the position, the timing between two robots needs to be synchronized [12]. The purpose of the synchronization is to make sure the robot passes the ball at the shooting area and at the same time, another robot will move with obstacle avoidance and then shoot the ball from the shooting area into the goal. From the test results, it is concluded that the positions of the robots have a high influence on the results.

Acknowledgements. The authors would like to thank Universiti Kebangsaan Malaysia for providing facilities and financial support under Arus Perdana Grant No. UKM-AP-ICT-17-2009, Fundamental Research Grant Scheme No. UKM-TT-03-FRGS0131- 2010 and UKM-PTS-2011-047.

References

1. Asada, M., Kitano, H.: Robotics and Autonomous System. The RoboCup Challenge, 3–12 (1999)
2. Kim, J.H., Shim, H.S.: A Cooperative Multi-Agent System and Real Time Application to Robot Soccer. In: IEEE Int. Conf. on Robotics and Automation, Albuquerque, pp. 638–643. IEEE Press, New York (1997)
3. Pratomo, A.H., Prabuwono, A.S., Abdullah, S.N.H.S., Zakaria, M.S., Nasrudin, M.F., Omar, K., Sahran, S., Nordin, M.J.: The Development of Ball Control Techniques for Robot Soccer Based on Predefined Scenarios. Journal of Applied Sciences 11(1), 111–117 (2011)
4. Kim, J.H., Lee, J.J., Fukuda, T.: Evolutionary Robotic System. In: Micro-Robot Soccer Tournament, pp. 29–37. KIST Publishing Company (1997)
5. Kim, H.R., Hwang, J.H., Kwon, D.S.: Co-operative Strategy for an Interactive Robot Soccer System by Reinforcement. Int. J. Control, Automation, and Systems, 236–242 (2003)
6. Egly, U., Novak, G., Weber, D.: Decision Making For Mirosot Soccer Playing Robots. In: 1st CLAWAR/EURON/IARP Workshop on Robots in Entertainment, Leisure and Hobby, pp. 69–72 (2005)
7. Huang, L.: Velocity Planning for a Mobile Robot to Track a Moving Target a Potential Field Approach. J. Robotics and Autonomous Systems 57, 55–63 (2009)
8. Liu, J.S., Liang, T.C., Lin, Y.A.: Realization of a Ball Passing Strategy for a Robot Soccer Game: a Case Study of Integrated Planning and Control. J. Robotica 22, 329–338 (2004)
9. Kim, J.H., Kim, D.H., Kim, Y.J., Seow, K.T.: Soccer Robotic. Springer, Heidelberg (2004)
10. Xue-Dong, H., Bing-Rong, H., Wei, M.: Distributed Control for Generating Arbitrary Formation of Multiple Robots. ROBOT 25(4), 66–72 (2003)
11. Alami, R., Ingrand, F., Qutub, S.: A Scheme for Coordinating Multi-Robot Planning Activities and Plans Execution. In: 13th European Conf. on Artificial Intelligence, pp. 617–621 (1998)
12. Pratomo, A.H., Prabuwono, A.S., Zakaria, M.S., Omar, K., Nordin, M.J., Sahran, S., Abdullah, S.N.H.S., Heryanto, A.: Position and Obstacle Avoidance Algorithm in Robot Soccer. J. Computer Science 6(2), 173–179 (2010)

Author Index